Contents

KU-750-140

Foreword

by Dr Elphis Christopher

It is a great honour to write the foreword to this book for women celebrating sixty-five years since the opening of the first birth control clinic by Marie Stopes. Marie Stopes, a formidable and energetic woman, wanted to free women from the unremitting burden of repeated pregnancies and gynaecological ill-health. Her book *Married Love* sought to save women (and men) from unhappy sexual relationships. She herself had experienced an unconsummated marriage; she read everything available on sex and the law relating to it and managed to get her marriage annulled.

In 1921, when Marie Stopes opened her clinic, maternal mortality was high, dangerous backstreet abortions were resorted to when pregnancies were unwanted and there were only two birth control methods – the sheath and cap. Today, as a result of better health care and nutrition, antibiotics and blood transfusions, the maternal death rate has been greatly reduced. Abortion, though not 'on demand', is legally available for any woman whose mental and physical health might be at risk from an unwanted pregnancy, and there are a variety of reliable birth control methods.

However, greater sexual freedom has brought other problems in its wake – an increase in abnormal cervical smears in younger women, a rise in sexually transmitted disease especially among women, leading to infertility and chronic gynaecological problems, and now the spectre of AIDS. The birth control methods, while not as risky as sometimes portrayed, nevertheless do have sometimes unpleasant and very occasionally life-threatening side-effects. Women themselves have higher expectations and are no longer prepared to take a 'back seat'. At the same time, relationships between the sexes appear to be more complex and more choices face women that are exciting but also challenging.

Women therefore need to be as well informed as they can be about their bodies and how they function so that they are better able to care properly for themselves and take active responsibility for their own sexual health. This book aims to provide women with up-to-date comprehensive information so that they can do just that. Armed with such knowledge, not only will the choices a woman makes for her health and sexual life be better informed, but she can also be more confident in her dealings with the medical profession. Marie Stopes would surely have approved!

Message from the Marie Stopes Clinic

by Kim Lavely

Those of us who work at Marie Stopes House are often asked why so many women are willing to pay for birth control and well-woman services here, when the same services are available without charge on the NHS. Apart from the issues of convenience and lack of delay in being seen, the fundamental reason is that, at 'Stopes', women are treated like intelligent human beings, given far more information, and encouraged to be actively involved in decisions concerning their health. Clients carry their own medical records and so know what is being written about them. The doctors and nurses do not wear white coats and are often on first-name terms with regular clients. These factors help to create an atmosphere in which clients feel they can ask questions without fear of a patronizing response.

This book was designed with much the same idea in mind. It has been written to provide you, your mother, sisters, friends and daughters with more and better information so that you can:

- take preventive action to protect and preserve your health
- respond quickly and appropriately to symptoms when they do occur
- be better prepared to interact with the medical profession when necessary, making realistic and well informed demands of the services available

And, although not written specifically for a male audience, this book could also help to educate lovers, husbands, brothers and sons about the health issues facing the women they care about.

Concerns about health problems, especially those affecting our sexual relationships, often give rise to tremendous anxiety, which can make the problems that much harder to cope with. *Your Body* aims to reduce those anxieties by increasing your sexual health awareness. It is intended as a reference book, so, although you may wish to read it cover to cover, you can just as easily dip in and out to learn more about a particular subject if you wish. Many topics are dealt with in more than one chapter, so do take the time to check each reference.

We welcome *Your Body* as a practical guide written by women for women. It may not answer all your questions, but it provides the tools you need to formulate the right questions and helps you know where to look for the answers.

1 *The Marie Stopes Clinic*

by Kim Lavely

'Dr Marie Stopes can fairly be said to have transformed the thoughts of her generation about the physical aspects of marriage and the rôle of contraception in married love.' This is what *The Times* had to say about Marie Stopes following her death in 1958, and, if anything, the paper underestimated the impact she would have. Women of every subsequent generation have had Marie Stopes to thank for increasing awareness of the importance of a healthy and enjoyable sex life for both men *and* women; and, in recognizing that the biological reality of a healthy sex life is the need for contraception, we owe her a debt of thanks for establishing the first birth control clinics in Britain.

Marie Stopes and the first clinics

Born in 1880, Marie was educated at home until the age of twelve. A highly ambitious woman, she was awarded first class honours in botany and geology from University College in 1902, and received her PhD in Munich in 1904. A year later she had become the youngest Doctor of Science in England, was appointed as a lecturer at Manchester University, and became highly regarded for her work in paleobotany.

She was married for the first time, at the age of thirty-one, to another scientist, Dr Reginald Ruggles Gates, but the marriage was not sexually successful, and in 1914 she had the marriage annulled on grounds of non-consummation. Her unhappiness with this marriage led her to write a book on marriage and sex (while still a virgin), emphasizing the importance for women as well as men of a happy sexual relationship. Published in 1918, *Married Love* created a sensation, selling 2,000 copies in a fortnight. The book had little information about birth control, but, in response to many letters from her readers asking for advice, Marie Stopes researched and published *Wise Parenthood,* a guide to those contraceptive methods available at the time.

Having recognized the need for wider availability of birth control for women of all classes, Marie Stopes and her second husband, Humphrey Roe, opened the first British birth control clinic in 1921 in Holloway; it moved in 1925 to its present location in Whitfield Street, near Tottenham

The cover of Marie Stopes's third book, published in 1920. *Right:* Marie Stopes with her son Harry in 1926. *Below:* The Mother's Clinic for Constructive Birth Control in Whitfield Street (now Marie Stopes House), where it moved in 1925.

Marie Stopes (seated) is shown here with a group of nurses at the Mother's Clinic in Holloway in the early 1920s. *Right:* The caravan Marie Stopes set up in 1928 to provide the north of England with birth control services. The caravan was later burned by a catholic fanatic.

Court Road in central London. Through a combination of numerous publications, public speaking engagements, legal battles, and clinic openings (including the first mobile clinic set up in 1928 to tour the north of England), Stopes brought contraception into the public arena, and made it possible for those who followed in the family planning field to achieve the wide availability that she thought so important.

The problem years

After her death in 1958, Marie Stopes's clinic activities were carried on by the women with whom she had worked, mainly by her secretary, Joan Windley. The Whitfield Street clinic, left in Stopes's will to the charity the Eugenics Society, became known as the Marie Stopes Memorial Clinic, and from 1958 until 1976 continued to provide a private but relatively low-cost family planning service independent of the Family Planning Association (FPA), which during this period operated a large and growing number of clinics.

By the mid 1970s, however, the FPA had achieved its ambition of handing over its thousand or so clinics to the NHS, to ensure that family planning was a free and integral part of the health service; within two

years general practitioners were brought into the scheme, greatly widening the number of outlets available for contraceptive services. These developments left the independent and fee-charging Stopes clinic with the difficult task of competing for patients, so finally the decision was taken by the Marie Stopes Memorial Foundation to go into voluntary liquidation.

The Eugenics Society, wishing to prevent the closure of the clinic and to see it continue to offer birth control services, sought the help of another British charity, Population Services. Population Services had until then worked only in developing countries, but accepted the challenge of rescuing the ailing London clinic, and assumed the management in early 1976.

The transition period

In order to avoid the financial pitfalls of the previous occupants, it was clear that the newly re-opened clinic would have to adopt a different approach from that of the traditional family planning clinic if it was to attract sufficient numbers to keep the doors open. This suited Population Services, a charity committed to providing innovative services designed to meet the changing needs of sexually active women (and men) wishing to avoid or delay pregnancy.

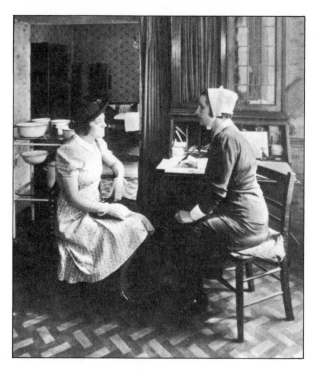

An early consultation: a nurse at the Mother's Clinic in Whitfield Street (now Marie Stopes House) talks to a patient.

The first step was to recognize that these are *clients,* not patients. As Dr Tim Black, founder of Population Services, writes, 'The majority of the fertile are certainly not sick, so the medical concepts of health care delivery are largely irrelevant to family planning.' *(Ten Institutional Obstacles to Advances in Family Planning,* by Timothy R. L. Black, from *New Concepts in Contraception,* edited by Malcolm Potts and Clive Wood, Medical and Technical Publishing Co Ltd, 1972.) After all, those attending the clinic were parting with hard-earned money when they could just as easily have gone to a free NHS clinic, so there had to be something special at Marie Stopes House, as it became known. Clinic services, opening times, and so on, were geared to the requirements of the women attending, not to those of the medical staff, as is so often the case in the provision of health care. Feedback from clients in the form of confidential questionnaires was regularly sought to find out how the clinic could improve, and the suggestions often put into practice.

The range of services at Marie Stopes House was broadened to incorporate other aspects of sexual health care. 'Well-woman screening', a concept almost unheard-of in Britain in the mid 1970s, was added, and continues to form an important part of the service, especially with the increased awareness of the need for regular cervical smears and breast checks.

In keeping with its commitment to making services more acceptable and less intimidating to clients, a simpler approach to female sterilization was pioneered in 1978, offering women an outpatient procedure with a choice of local or general anaesthetic, enabling them to be back in their homes within a couple of hours after the operation. While still not quite as simple as vasectomy (by virtue of being an abdominal operation), it is a great improvement over the three-day stay in hospital and obligatory general anaesthetic that are still the norm in most hospitals.

And not to forget the men, 'Stopes' has developed a five-minute vasectomy technique that takes a lot of the fear and fuss out of male sterilization. At the request of Stopes clients, a same-day counselling and operation service is provided for those men and women who, having made a firm and final decision to be sterilized, wish to minimize the delay.

Services are no longer limited to the Whitfield Street clinic; there are now Marie Stopes Centres in Leeds and Manchester, and Stopes Vasectomy Centres in eighteen towns throughout Britain.

The Stopes legacy

We know that some of the services offered would not have met with Marie Stopes's approval. For example, the Marie Stopes Annexe in Whitfield Street sees thousands of women each year for advice and help with unwanted pregnancy, and the Marie Stopes Nursing Home in north London offers abortion on a daycare or overnight stay basis, again with a choice of anaesthetic. Marie Stopes, although a strong advocate of birth control and of women having control over their own bodies, was totally opposed to abortion, not on moral grounds but because in her time abortion was not the safe procedure it is now, and was often performed by unqualified practitioners outside the law.

The range of contraceptive methods was nowhere near as great in the 'Stopes era', and she felt strongly about which methods should and should not be used. She wholeheartedly advocated the use of the cervical cap with a quinine pessary, and, although she mentioned most other available methods in her book *Wise Parenthood,* she discouraged their use on the grounds that they were unsafe or detrimental to an enjoyable sex life. Certainly many couples would agree with her even today.

Marie Stopes was far in advance of her time not only in advocating the use of contraception but in the way she chose to provide her clinic services. She believed that women attending her clinics would find it easier to establish a rapport with trained nurses than with doctors, and so most visits were with qualified midwives. Today the concept of paramedical provision of services is widely discussed, and, at Marie Stopes House, many clients see nurse practitioners for birth control or well-woman services.

The young Marie Stopes.

Postscript

Writing about the activities of the Marie Stopes Clinic several years after his mother's death, Dr Harry Stopes-Roe said, 'The services provided ... are doing today what is called for today, just as Marie Stopes herself did what was called for fifty years ago. The interesting thing is that Marie Stopes would, if she were alive, surely disapprove of what is being done now. This, I think, shows something of the nature of real pioneering.' *(Marie Stopes and Birth Control,* H.V. Stopes-Roe with Ian Scott, Prior Press Ltd, 1974.)

We at Marie Stopes hope that, although she might well disapprove of what goes on under 'her' clinic's roof, Marie Stopes would recognize that the services offered today are improving the quality of life for those women and men who make use of them, by helping them achieve healthier more enjoyable sexual relationships without fear of unwanted births.

2 *Your Body*

by Penny Chorlton

The changes that mark the onset of maturity in a young woman occur at puberty. The actual age can be anything from as young as nine to as old as seventeen, but for most girls the physical changes start at the age of thirteen, and signal the beginning of the reproductive years.

The first signs are the development of breasts, followed by the growth of pubic and underarm hair and the commencement of menstruation, or periods. Understandably, late developers can become very self-conscious about their lack of any of these outward signs of maturity and fear, usually entirely without justification, that they are abnormal.

Menstruation

The timing of a girl's first period is usually influenced by several factors. One of these is heredity – a girl is likely to start her periods around the same age that her mother started hers. However, it is unlikely that she will start at the same age as her grandmother because research shows that in the late nineteenth century and for the first few decades of this century, the average age of onset of menstruation was sixteen.

One of the reasons believed to be behind this is diet. Women in the past were not as well-nourished, and were slightly smaller than women today. Diet and nutrition are still important factors in marking the development of young women. Taller, better-fed, heavier girls tend to start their periods before their slimmer, smaller, less well-nourished peers. Girls who diet fanatically at this age may also delay the physical changes that mark the beginning of becoming a woman capable of reproduction.

The time between periods – that is, the length of the menstrual cycle – can be anything from twenty-one to thirty-five days, with an average of twenty-eight days. The interval between periods, even if it varies from time to time, does not matter as long as there is some sort of regular recurring pattern.

The average woman will lose up to half a teacup of blood in each period. Hygienic collection of the menstrual flow is purely a matter of personal choice, and while older women often prefer the idea of sanitary towels used externally, younger women, especially active sporty types, usually prefer tampons which are inserted into the vagina and conceal any sign that they are menstruating. The added advantage of collecting the blood internally is that there is less risk of leakage and messy and embarrassing staining. It is not necessary for intercourse to have taken place for young girls to wear tampons. Even if the hymen is intact (and it can be broken by activities other than sexual intercourse) a tampon can be safely worn.

The rare condition known as toxic shock syndrome has been linked with tampon use. However, the association is unclear and should not cause undue concern. If a woman suddenly develops symptoms such as a high temperature, headache, vomiting, diarrhoea and a rash, she should remove the tampon and consult her GP immediately.

During the first ten days of the cycle, the part of the brain called the hypothalamus produces hormones that act on the front part of the pituitary gland, which in turn produces two substances: the follicle-stimulating hormone FSH, and the luteinizing hormone LH. The FSH stimulates the ovary to release an egg and the hormone oestrogen is released.

This phase of the menstrual cycle is unnoticeable to the woman herself. At the same time, the womb starts preparing itself in case a fertilized egg arrives, with the help of the hormone progesterone – literally, the word means 'pregnancy-making' – which builds up as the oestrogen levels subside. Progesterone makes the lining of the uterus, the endometrium, thicken and develop extra blood vessels which are then ready to supply food and oxygen to the fertilized egg should it arrive.

The second phase of the cycle is when ovulation takes place, usually around day twelve, thirteen or fourteen after the first day of menstrual bleeding. This is when oestrogen production is at its peak and so is a woman's fertility. Sexual intercourse at this time is most likely to result in pregnancy and is therefore to be avoided if pregnancy is undesirable and maximized if it is wanted.

The second half of the cycle will vary according to whether fertilization has taken place; if it has not, the lining of the womb will start to shed round about day twenty-eight. This leaves by a tiny hole in the cervix and passes through the vagina and out by the vulva, to appear as a period.

For the vast majority of women, menstruation should be seen as normal and not an illness. All the usual activities can be pursued, including sports and bathing. The theory that you should not bath or wash your hair is foolish – indeed, the cells that produce sebum can become more active and make your skin and hair greasier than usual, in which case more frequent washing may perhaps be necessary.

Dysmenorrhoea
Some girls are unfortunate and suffer from painful periods (dysmenorrhoea). Part of this *may* be psychosomatic, but, in a small proportion of cases, girls

do genuinely have a bad time because their periods are heavy, with substantial blood loss, or they are irregular, or because they suffer cramps and severe lower abdominal pain. Dysmenorrhoea can be treated and girls who are suffering serious discomfort which interferes with their normal day to day activities should consult a GP.

One of the most common forms of treatment is to prescribe the contraceptive pill, because this reduces the impact of normal menstrual cycles by allowing artificial hormones to suppress the natural ones in the body responsible for causing the monthly physical changes that are so unbearable. Painkillers can also be helpful, and healthier drug-free methods of coping with painful periods are a viable alternative preferred by many women (see Chapter 4 and Bibliography).

The reproductive system

The reproductive system in women consists of breasts, the womb or uterus, ovaries, fallopian tubes, and vagina. Each month, one of the ovaries produces a ripe egg, which is despatched along its fallopian tube to the womb. The female sexual organs also include, internally, the cervix and, externally, the vulva, the appearance of which varies from woman to woman.

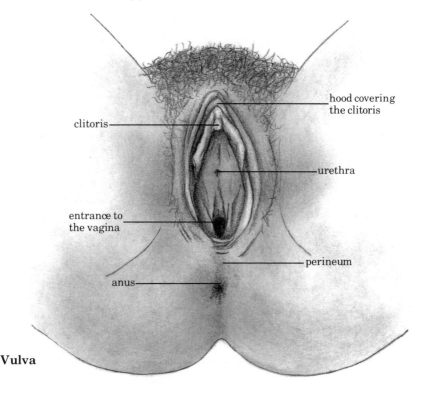

clitoris

hood covering the clitoris

urethra

entrance to the vagina

perineum

anus

Vulva

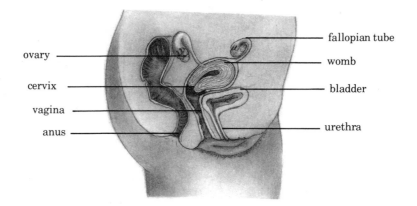

The vagina

This is the passage from outside the body leading to the womb and is about 10 to 12cm long in the fully-grown woman. Because it is made up of muscle tissue, normally the walls of the vagina lie close together but during intercourse they stretch to accommodate the penis, while during childbirth they expand greatly to allow a baby to be born.

The vagina secretes a moist fluid which lubricates it and makes intercourse more comfortable as well as being a natural method of keeping the area clean and free of infection. After the menopause, the hormones in the body stop producing this moisture and the vagina becomes more vulnerable to infection (see Chapter 10). Douches and deodorants interfere with this natural cleansing process and should be avoided.

The vulva

This is the external part of the female genitalia, and contains the entrances to the vagina and the urethra. The urethra is the end of the tube leading to the bladder and is where urine emerges. Just inside the vulva is the clitoris, and, despite Freud and his followers' comments to the contrary, this is the sensitive centre of a woman's erotic sensations.

Careful washing in this area is important and can help avoid infections. The urethra in women is much shorter than in men so it is relatively easy for germs to enter and infect the bladder, thus causing cystitis. The anus is also very near the vagina and the urethra and it is not difficult for fecal germs to travel and cause infection. Regular washing with warm water and mild soap is therefore important, as is washing or wiping *away* from the vulva after defecation.

The womb

The womb, or uterus, is about 9cm long and 6cm across its widest part, but expands to thirty times its basic shape and weight during pregnancy.

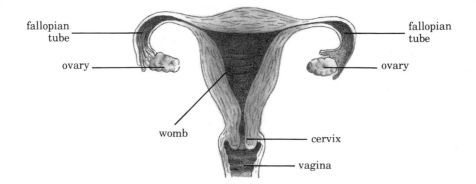

At the lower end is the neck or cervix, which projects about 2cm into the vagina. This makes examination of the cervix very simple and, with practice, any woman can explore her accessible internal sexual organs with a clean finger. If she detects any abnormalities, or suffers any unusual symptoms, a GP can make a fairly thorough basic investigation. Once a woman starts having intercourse, she should present herself for regular cervical smear tests (see Chapter 13).

The cervix feels soft and round and has a small dimple in the centre (the cervical os), through which the menstrual blood is shed. It seems incredible, but this tiny entrance, about a millimetre wide, can expand during pregnancy to let the baby through. Thick mucus deters any sperm from entering the womb during the infertile phase of a woman's menstrual cycle but during ovulation the mucus thins out.

The rest of the womb lies forward at an angle of about ninety degrees to the vagina. The walls are made of thick stretchy muscle with a hollow cavity, at the top of which are connected the fallopian tubes.

The fallopian tubes

There are two tubes, connected at either side of the top of the womb, which form an arch and end in a funnel-shaped finger-like arrangement that surrounds the ovaries. The tubes are lined with tiny hairs which help waft the egg towards the womb. They are also stocked with nourishing substances which the egg can feed on if and when conception occurs.

Fertilization, if it is to take place, can only occur within the fallopian tubes. If this happens, helped by the tiny hairs, the fertilized egg is transported along to the womb, where the cosy lining has been made ready. Sometimes, the fertilized egg gets stuck in the fallopian tube, or elsewhere outside the uterus, causing an ectopic 'misplaced' pregnancy. Such a pregnancy is doomed to failure and, unless diagnosed and treated, can lead to serious damage of the reproductive organs. Usually the whole tube is removed, but this does not mean future pregnancies are impossible as one healthy fallopian tube is quite sufficient to make fertilization possible. Fortunately, the risk of a second ectopic pregnancy occurring in the remaining fallopian tube is very small.

A pregnancy that is allowed to continue in the fallopian tube eventually strains it so much that it bursts into the pelvic cavity. Babies have been known in rare cases to fasten onto a section of gut and form a placenta and manage to achieve a normal life through this extraordinary means.

The ovaries

The ovaries are the female egg cells and are the equivalent of the testes in the male, which produce the sperm. These two endocrine glands are located on either side of the uterus, and have two functions: to produce the hormone oestrogen and to release an egg every month throughout a woman's reproductive life (which lasts approximately thirty to forty years).

In a fully-grown woman, the ovaries measure about 2cm wide and 3–4cm long. After the menopause, they shrink to half that size. The mature ovaries are filled with follicles, which are tiny fluid-filled structures containing egg cells. The ovaries are enclosed in a very fine membrane which can sometimes swell and form cysts. These cysts can enlarge and cause abdominal swelling, in which case they may have to be surgically removed (see Chapter 13).

The breasts

The breasts contain milk glands, pectoral muscles, lymph nodes and fat. They undergo dramatic changes throughout a woman's reproductive life, changing during each menstrual cycle in readiness for a possible pregnancy and enlarging during pregnancy itself.

The milk channels develop from the nipple and divide into smaller and smaller branches. During pregnancy, milk production occurs quite early and can continue throughout pregnancy, although, of course, it is not required at this stage. Breasts enlarge during breast-feeding and shrink back again several weeks after feeding has ceased.

The female form

It is quite astonishing that people, particularly women, are so obsessed with trying to remain slim. Indeed, at any one time, eleven million people – and two thirds of all women – are on some kind of diet, or between diets. From vacuuming out the fat via plastic surgery, to inserting balloons in the stomach to wiring the jaws, there seems no end to the extraordinary lengths to which people will go both to invent – and use – slimming devices. Conventional wisdom has it that for most people the solution to being overweight is quite simple – take more exercise (burn up more calories) and eat less (take in fewer calories).

A new theory has recently been developed that may explain why so many modern diets do not work. 'Persistent Fat Retention' (PFR) is the term used to describe the problem of those who cannot shed fat no matter what they do. Its advocates claim that people who suffer from PFR apparently do not lose fat when they go on a diet; something in the body signals that the fat stores *must* be kept, come what may, and so the hapless dieter sheds lean muscle instead, even the tissue from vital organs like the heart, liver and kidneys. This is why dieting can be dangerous for certain people. For them, it has to be discovered *why* their bodies are determined to hang onto their excess fat, and then the source of the problem should be tackled, not the symptoms of it.

PFR was discovered by authors Arabella Melville and Colin Johnson through research that was published in 1986 in their book *Persistent Fat and How to Lose It*. They concluded that PFR is a system by which the body stores fat as a protection against poisons, which can be anything in the environment or a reaction to taking large and regular amounts of drugs such as tranquillizers, anti-inflammatories and the Pill.

Many drugs such as the Pill actually state on the label that a possible side-effect may be weight gain but it has always been assumed that such side-effects would disappear once the drugs were stopped. However, it may be that the bodies of such people continue to keep hold of the protective fat just in case another 'attack' of drugs is administered. For example, women who go on the Pill frequently put on weight in the order of half a stone; when they stop, they rarely lose the same amount. If the theory is correct, then by reducing exposure to poisons – in food, drugs, alcohol, smoking and the environment – fat reserves *may* shrink.

Doctors who follow conventional thinking assume that people grow fat because they have houses full of labour-saving devices, sit at home watching TV every evening, use cars instead of walking and lead generally inactive lives at home and at work. Based on this theory, there are innumerable books, magazines, articles, and 'magic-formula' products marketed offering various short cuts on the calorie-reduction merry-go-round. The diet-merchants are always reappearing to sell the next and latest 'breakthrough' slimming aid.

Yet it is an unexpected, very unfair fact that some big eaters can stay thin while some small eaters still get fat. This is because metabolic rates vary tremendously; some people burn off surplus energy quickly while others use it sparingly and store the rest as fat. This constant struggle against nature's decree that people should be a certain size can cause untold misery and physical and psychological damage, which in its extreme form may manifest itself as either anorexia or bulimia.

Unfortunately for the naturally plump, ever since Twiggy burst on the fashion scene in the 1960s millions of women in Britain and the rest of the Western world have tortured themselves with the unreasonable desire to conform to a supposed ideal size regardless of their individual natural body structure. Especially since the 1970s, millions of women have become obsessed with 'correcting' the shape they acquire at birth and often hate their natural form. Unfortunately, if it is in a person's genes to be on the large and big-boned side, no amount of slimming will make that individual sylphlike. Any woman thinking of losing weight would be well advised to study her female relatives first.

Recent surveys show that a third of women (thirty-two per cent) are overweight, and eight per cent obese, while thirty-nine per cent of men are overweight and six per cent obese. In the over sixties age group, fifty per cent of women and fifty-four per cent of men are overweight.

Doctors have realized that the dangers of obesity per se have perhaps been exaggerated in the past and now slimming priestesses like Jane Fonda are saying that it is all right to be fat, as long as you are fit. Perennial dieters should take note: womanly curves are quite acceptable, but unhealthy flab is not, and that can usually be worked off without fancy diets. Exponents of this new fitness look positively encourage women to have feminine curves, as long as these are well-toned muscle, not flab. The

Suggested acceptable weight ranges

Height without shoes	Weight without clothes	Height without shoes	Weight without clothes
1.45m 4ft 9in	**42kg – 50kg** 6st 9lb – 7st 12lb	**1.68m** 5ft 6in	**56.5kg – 63.5kg** 8st 12lb – 10st 0lb
1.47m 4ft 10in	**45.5kg – 51.5kg** 7st 2lb – 8st 1lb	**1.70m** 5ft 7in	**57.5kg – 65kg** 9st 1lb – 10st 3lb
1.50m 4ft 11in	**47kg – 52.5kg** 7st 5lb — 8st 4lb	**1.73m** 5ft 8in	**59kg – 67kg** 9st 4lb – 10st 7lb
1.52m 5ft 0in	**47.5kg – 54kg** 7st 7lb – 8st 7lb	**1.75m** 5ft 9in	**61kg – 68kg** 9st 8lb – 10st 10lb
1.55m 5ft 1in	**49.5kg – 55.5kg** 7st 11lb – 8st 10lb	**1.78m** 5ft 10in	**62.5kg – 70.5kg** 9st 11lb – 11st 1lb
1.57m 5ft 2in	**51kg – 57kg** 8st 0lb – 8st 13lb	**1.80m** 5ft 11in	**64kg – 72.5kg** 10st 1lb – 11st 5lb
1.60m 5ft 3in	**52kg – 58kg** 8st 2lb – 9st 2lb	**1.83m** 6ft 0in	**66kg – 74kg** 10st 5lb – 11st 9lb
1.63m 5ft 4in	**53kg – 60kg** 8st 5lb – 9st 6lb	**1.85m** 6ft 1in	**67.5kg – 76kg** 10st 9lb – 11st 13lb
1.65m 5ft 5in	**54kg – 62kg** 8st 7lb – 9st 10lb	**1.88m** 6ft 2in	**69.5kg – 77.5kg** 10st 13lb – 12st 3lb

sad thing is that despite the efforts of the huge multi-million pound slimming industry, very few overweight people manage to reduce their weight dramatically and keep it off permanently. Despite the promises of the 'before' and 'after' photographs, few people, even the subjects of those photos themselves, are ever 'cured' of fatness, often because their goals are unrealistic in terms of their anatomy and the size of their bones.

According to the height and weight charts, many people who are permanently on a diet are actually of quite acceptable proportions, but for various reasons some of them prefer to dismiss the validity of the charts rather than question their own attitudes and self-images. For many women, the psychological consequence of calorie-counting is that they end up eating far more than usual because food assumes enormous significance in their lives. This often leads to the bingeing and fasting see-saw that characterizes the lives of so many perennial dieters.

Having said all that, it is obviously important not to get overweight if you can possibly avoid it, as most of us can, and preventing a middle-age spread or pot-belly forming in the first place is much easier than trying to get rid of one later. As well as being unattractive and unfashionable, obesity can lead to various potentially serious medical problems such as hypertension (high blood-pressure) as well as other heart and kidney problems, diabetes, gallstones and a number of cancers. The diseases associated with being overweight are also caused by smoking and a lack of exercise so it is important to consider your whole lifestyle when trying to lose weight, and not merely food intake.

Anorexia nervosa

Anorexia means loss of appetite and this is often self-induced in people who desperately want to be thin. Anorexia nervosa has thus become known as the 'slimmer's disease'.

Ninety-nine per cent of anorexics are female and most of these are adolescents, but recent surveys have shown that boys can develop anorexia and that two per cent of all women are anorexic and many more are of an anorexic disposition.

Anorexia is fundamentally a psychological problem and doctors have found that anorexics have usually suffered some kind of psychological disturbance. Their self-image becomes distorted and they lose perception of their real body-shape and do not accept the judgement of others – including doctors – that they are becoming skeletal and dangerously thin. Frequently they continue to describe themselves as being fat when any objective observer would describe them as painfully thin.

Their reasons for desiring to be thin can vary from wanting to obtain control over their bodies to wanting to be weak and looked after. Many

Sufferers of anorexia nervosa often have a distorted image of their bodies. This patient has expressed her feelings in a drawing. The larger sketch is called 'How I feel inside' and the smaller one 'How I would like to be'.

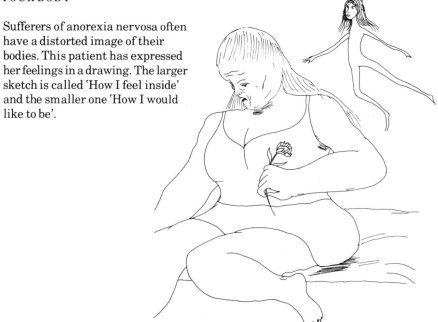

young girls suffering from this disease are often trying to stave off womanhood by preventing the sort of curves that men find sexually erotic. This type of victim – by far the most common sort – is using her body both as a means of self-expression and as an object of her own will. Some research shows that if such a girl starts feeling anxious about the biological changes that are suddenly happening to her, and she has a background of parental inhibition, anorexia can result, especially if she does not get on very well with her mother.

Older women can become anorexic after a successful bout of slimming. They live on near-starvation diets, their stomachs shrink, and they are almost certainly being truthful when they say they do not feel hungry, since they have conditioned their bodies not to expect food. Anorexics usually get their weight down to below seven stones.

Psychological counselling may be necessary to help the sufferer get things into perspective again and, in severe cases, hospitalization and force-feeding may be necessary to prevent death. Serious cases need psychiatric help to convince them that their self-image has become distorted to the point where their lives are in danger. Often a girl or woman will only consult a GP about her condition if she has become concerned about her lack of periods, or the growth of extra body hair or some other symptom, rather than the weight loss itself.

Recent research has suggested the possibility that anorexia nervosa could have a physical cause and one theory put forward is that the victims could be suffering from zinc deficiency in their diet. Since zinc is known to deplete the senses of taste and smell, as well as cause mental depression – all of which are symptoms suffered by anorexics – this theory is well worth exploring.

The outward symptoms of anorexia
- Hyperactivity
- Loss of periods as hormonal activity slows down
- Loss of head hair
- Dental problems due to lack of calcium
- Fevers and other biological upsets resulting in high and low mood swings
- The appearance of downy hair on the body

Bulimia nervosa

Bulimia nervosa is an eating disorder that has only come to light in the last few years, although it has probably existed for some time. The sufferer continues to eat, but is so scared of putting on weight that she either makes herself vomit, or takes laxatives to expel the meal quickly. Some people who have to keep their weight down for professional reasons, such as jockeys, ballet dancers and models, have also been known to use this 'technique' as a method of weight control.

During 'bingeing', the amount of food consumed can be enormous, literally thousands of calories. This is because the bulimia sufferer knows she can get rid of the calories by inducing vomiting. Thus, like the anorexic, bulimics are exerting control over their bodies.

Sufferers are harder to identify than anorexics because to friends and family they appear to continue to eat normally and their body weight does not decrease as dramatically as it does with anorexia nervosa. However, the acid produced by repeated vomiting is extremely bad for the digestive system, and can eventually harm the enamel on the teeth.

Because bulimia is so easy to conceal, nobody knows how many sufferers there are. However, a questionnaire published in a major women's magazine elicited over 1,000 replies, almost exclusively from women. Over half of them said they made themselves vomit at least once a day. Significantly, although most of the women admitted they needed medical help, less than a third had mentioned their habitual vomiting to a GP.

3 *Sexual Awareness*

by Dr Elphis Christopher

It was a man, Freud, who asked the question what is it that a woman wants. The sex researchers in the United States, Masters (a man) and Johnson (a woman), investigated what happened when a woman responded sexually though they did not ask what women felt at the time. Marie Stopes herself was in no doubt – women wanted to be 'wooed', gently caressed to sexual arousal. In *Married Love,* first published in 1918, Marie Stopes describes an ignorant and foolish husband who did not know that 'a husband's lips upon her breast melt a wife to tenderness', and that it is one of 'the surest ways to make her physically ready for complete union'. The language may seem quaint and archaic to modern ears and yet the meaning is clear – a happy sex life takes time and trouble.

Sadly, women have often not made it known to their partners what they want, perhaps because of age-old feelings that women are not supposed to be much interested in sex and that men know what to do (a view promoted by men). There was, and probably still is, the wish to protect male vanity. Women, also, have been fearful or ashamed of finding out for themselves what they like and want. However, in these post-pill times women are supposed to have become sexually rampant and multi-orgasmic, making their men impotent in the face of their insatiable sexual desires. It is all very confusing and bewildering for both sexes.

The evidence from clinics dealing with sexual problems, where women tend to present more often than men, is that sexual difficulties and dissatisfactions are very common. This may be because women are no longer prepared to put up with an unsatisfactory sexual relationship but also because it is easier nowadays to admit to a sexual problem. Although it might be thought that most of these problems have a physical cause, the majority have emotional roots or may depend on the kind of relationship the woman has with her partner.

The sexual life of a woman does not begin with the first time she actually has sexual intercourse, important as that is. What she feels about her body, her genitals and the whole process of growing up and being a woman will strongly influence her sexuality. A woman, unlike a man, has many bodily changes to cope with in her lifetime. The way in which each physical milestone, together with the feelings related to it, is handled, both by the woman herself and those around her, will have a profound effect on her sexual self-confidence and ultimately her sexual relationships.

Menstruation

The first important milestone is the onset of menstruation. In the past this was often treated in a cursory and dismissive or overanxious way. Many older women have related how their first period came as a shock. They were afraid to tell their mothers. They thought that they would bleed to death or that they had done something terrible to themselves and would be punished. They were taught that it was a woman's lot – the curse. The multitude of euphemisms given to periods shows how embarrassing and shameful menstruation was considered, instead of an event to be celebrated. Hopefully, today's woman will teach her daughters to be proud of their bodily functions. This, in turn, may lead to fewer menstrual problems.

Using tampons at the time of menstruation can be very useful for women in familiarizing themselves with their bodies. For women who are unable to get the tampon in the vagina it might be helpful to seek advice from a woman doctor. Family planning clinics, which are run mainly by women nurses and doctors, can teach the woman how to insert the tampon. It may be that the hymen (the thin, half-moon shaped piece of tissue) covering part of the vaginal opening will need to be stretched. The woman can be shown how to do this by inserting one then two fingers into the vagina. Some women are squeamish about doing this especially when the teaching from mother was that it is wrong to touch yourself. Being unable to fit a tampon often gives rise to anxiety that there is a 'blockage' inside the vagina and that intercourse will be extremely painful. This then sets the scene for a sexual problem, with fears of being ripped and torn, of being too small for intercourse. This, in turn, leads to a tightening of the muscles in the lower third of the vagina causing a condition known as 'vaginismus' so that intercourse is made difficult and uncomfortable, thereby confirming the woman's worst fears. Explanation and teaching how to stretch the vaginal opening can prevent this happening.

First intercourse

This is the second important milestone. It can be approached in a variety of ways by both sexes. For men there is the anxiety about whether they will actually be able to 'do it', that is, get an erection and maintain it sufficient time to enter the vagina. For women there may be a fear of pain and bleeding if the hymen has not been stretched. There may also be a fear of pregnancy; all too often, young people are not prepared contraceptively for this first act of intercourse. There may also be quite unrealistic expectations that the experience will be as wonderful as the romantic novelists describe it. More often than not it will be a letdown, with the man ejaculating quickly leaving the woman wondering what all the fuss was about. There will also be a sense of strangeness – of having something inside the body that may take some getting used to.

Other difficulties can relate to uncertainty the woman has about whether she really wants to be involved in a sexual relationship. She needs to be clear about her motives. Is she involved because she wants to keep up with her friends or because she is afraid she will lose her partner or be called names and told she is frigid? In the past, women were expected to be the guardians of sexual morality and to determine how far the man should go. She could rely on the fear of pregnancy to control the man to some degree. Now with the effective contraception provided by the Pill she does not have that excuse to fall back on. She needs more courage and self-confidence to say no. Robbie Burns, the Scots poet, has said, 'A standing cock hath nae conscience'. But what goes up can come down and the woman should not feel obliged to have sex when she does not want it!

After the first child

The first pregnancy and childbirth (whether intended or not) is the third important milestone. The body changes, the vagina is stretched and it may have had to be cut (an episiotomy). The delivery of the baby may have required forceps. Even with a normal pregnancy and childbirth the woman may have the most natural fear that she will die. It is only in the last fifty years that maternal mortality has dropped to very low levels (eleven per 100,000 births). After childbirth women often feel that their body, especially the vagina, does not belong to them – it has become 'medicalized' and no longer sexual. There may be a fear that they have been changed or damaged internally by the birth and this can lead to a loss of interest in sex. Realistically, sexual intercourse in the first few weeks after delivery can be uncomfortable due to vaginal dryness and even painful if there is an episiotomy scar. A lubricant such as KY jelly can be useful and lovemaking should be gentle at this time.

In an effort to have less medical interference in childbirth there has been an emphasis on 'natural childbirth'. Where this is successful women have a marvellous sense of achievement. However, it may not happen with every birth and women in this situation often feel that they have failed. They feel cheated and may become depressed. Sometimes partners are blamed for not giving enough support. The end result may be a sexual problem with the woman losing interest in sex, blaming herself and/or her partner. It is important for women to know that not all births can be 'natural' and that it is not their fault if the birth does not go according to plan.

Becoming a mother also involves emotional changes. It may alter the way a woman looks at life and her relationships. It may change her sexual feelings. Some women become sexually inhibited not just because of physical reasons but because of vague guilty feelings that mothers should not be sexual. This appears to relate to their feelings about their own mothers and the difficulty of accepting them as sexual people. Realistically, the emotional as well as physical demands on the woman with a new baby often result in tiredness and even exhaustion, and the consequent loss of interest in sex, usually temporary, at this time is quite normal.

The menopause

This is the fourth physical milestone for a woman. It marks the end of her reproductive life. The periods finish. There is a gradual reduction in the female hormones, notably oestrogen, that cause bodily changes. The vagina may gradually get smaller and the woman may experience dryness and consequent soreness with intercourse which may put her off sex. Failure to have sex can then exacerbate the problem. The Americans have a very succinct phrase for sexual activity at this time, 'Use it or you lose it'. This means that regular intercourse will help prevent the vagina shrinking and getting tight. Where soreness and dryness are a problem, hormone replacement therapy in the form of tablets taken orally or cream put into the vagina can be beneficial. Both sexes will find that as they get older, while sex can be as good as it ever was, sexual arousal may be slower and more care and time will be needed. Again, this is normal and should not be a cause for concern, but some people worry that because they are not the same at fifty-five as they were at twenty-five sex has to be given up. For some women, especially those who have been physically attractive in their youth, growing old can be a terrifying experience and there can be a desperation to hang on to youth – the mutton dressed up as lamb phenomenon. Growing old gracefully and healthily is an art that needs to be cultivated by attention to diet and regular physical exercise.

Loss of interest in sex

Some women grow up not interested in sex at all. This is usually due to the negative attitudes of parents about sex, which is seen as dirty or frightening. Some mothers are fearful of their daughter's sexual development, seeing it as a threat, and try to keep her like a child. The girl growing up in this environment feels alarmed and disgusted by her awakening sexual feelings and does her best to suppress them. The woman may not see this as a problem until she is involved in a sexual relationship and the partner gets fed up and frustrated by her lack of response. It then may become a battle of wills with her claiming that he does not really love her and only wants her for sex, whereas for him sex is part of a loving relationship. Sadly, in the past and in some cultural and religious groups, not being interested in sex was and is synonymous with being a 'good' woman. Such attitudes can be changed with specialist help provided the woman genuinely wants to change. Sometimes the mother's bad gynaecological or childbirth experiences are presented in such a way as to frighten her and put her off sex. Again, this may need exploration with a specialist in sexual problems.

More often, perhaps, the woman may have once been interested in sex but then lost any desire for it. The causes can be many and varied. It may be that the woman had unrealistic expectations of sex, that it did not require any effort from her. Disappointed that it did not turn out to be the exciting experience she expected, she then switches off. In these situations, there is usually poor sexual communication between the partners who neither know nor dare to find out what the other likes and wants. It is somehow assumed that to do so will imply criticism and risks rejection so both partners keep quiet.

Loss of interest in sex can also happen after a physical event such as childbirth or the menopause. Perhaps the most common causes, however, are to do with the relationship itself. The woman may feel hurt, angry and upset with her partner over any number of issues – money, the care of the children, in-law trouble, feeling neglected. The resentment builds up and she switches off sex. A common picture is of the woman stuck at home with the children, bored and lonely; the husband or partner comes home, has his meal then slumps in a chair, falling asleep in front of the TV, wakes up refreshed in time for going to bed and then wants sex. Meanwhile, the woman, feeling shut out and taken for granted, understandably cannot and will not respond to his sexual advances. A row then ensues and/or both partners sulk, thereby setting up a vicious circle. Sometimes the relationship has changed so much that the woman really wants to end it but is afraid of hurting the partner. This again can lead to a loss of interest in sex. The couple then think that they have a sexual problem whereas, in reality, it is the relationship itself that is falling apart. Help may be needed for the couple to face this together.

Painful intercourse

Complaints such as soreness, dryness and painful intercourse may have a physical cause that requires medical treatment. More commonly, emotional factors are involved whereby the woman is not becoming sexually aroused. The vagina fails to lubricate and enlarge, hence the soreness and dryness. The complaints may prove a legitimate excuse for avoiding sex where the woman cannot tell her partner that she does not want it. In these situations the partner may put pressure on the woman to get help. When she is examined, of course nothing much wrong will be found except, perhaps, slight spasm of the vaginal muscles.

Problems in the relationship causing loss of interest in sex can produce painful intercourse. Sometimes guilt and regret over an abortion, especially where the woman really did not want it but had it done for the partner or to save the relationship, can cause her to lose interest in sex or find intercourse painful. It is as if she has to punish herself and her partner. Often after an abortion the partner does not want to discuss it; the woman is then left with painful, sad and unresolved feelings. These may then get expressed in the sexual relationship, which may lead eventually to the break up of the relationship – something neither partner really wanted. It is important that such feelings are shared between the couple, not in a blaming, accusatory way, but helping each other to acknowledge and accept the sadness involved.

Unfaithfulness by the woman's partner, especially where this has resulted in the woman getting a sexually transmitted disease, can cause sexual difficulties even when the partner has regretted it and the woman thinks that she has forgiven him. The niggling doubts and uncertainties about whether she can ever really trust him again may lead her to withdraw sexually and be unable to get aroused.

Contraceptive problems, especially where the woman resents having to be the one who takes precautions and where the decision about which method to use has not been a joint one, can also lead to sexual difficulties. This will have to be brought out into the open and discussed properly. Some couples decide that they do not want children and the sexual relationship is happy and satisfying. As time passes, however, it can happen that one partner changes his or her mind and very much wants a child and tries to pressurize the other partner. Again, this tension can result in sexual difficulties and even a withdrawal from sex. (See also Chapter 13.)

Non-consummation

This means that the penis has been unable to penetrate the vagina. This problem is more common than popularly supposed. On the woman's part, she may be interested in sex and even be able to have orgasms with clitoral stimulation but there is intense fear of being penetrated. All kinds of myths and fantasies may surround this fear. There may be a fantasy that the vagina is too small (and the penis too big) and that it will be torn. The woman may believe she will bleed to death or that the penis will go into the wrong place. Some women believe that they have only one opening down below and that they may urinate or defecate when the penis enters. For some women the fear is really about pregnancy and childbirth, bound up with a reluctance to grow up and be a woman, with all that might entail.

For others intercourse represents submission to a man and means putting herself in his power. The vagina, hidden, unknown and private as it is, can be seen as a weak, vulnerable place through which the woman can be harmed. These fears and fantasies need exploring with help from a psycho-sexual counsellor or doctor, together with a vaginal examination carried out by the woman herself under the guidance of the counsellor to overcome the fears.

Orgasmic dysfunction

Ever since the work of Masters and Johnson looking at the male and female sexual response cycle, women who have not experienced orgasm have felt anxious that they were missing out. In some instances, this turned women off sex altogether; in others, women struggled to have an orgasm and ended up frustrated and angry. Some men feel that they have to 'give' their partners orgasms and failure to do so triggers fears that their penises are too small. Sometimes this has led men to accuse their partners of not being normal or lacking vital parts, which lowers the woman's sexual confidence even further. Women show enormous variation where orgasm is concerned, from those who will climax on most sexual occasions to those who experience orgasm only now and then. There are also women who never get an orgasm but who enjoy lovemaking and sexual intercourse and for whom not having a climax is only a problem when they feel they *ought* to have one. Some women learn to masturbate by rubbing around the clitoral area when they are girls and get a climax easily and quickly. If they are made to feel ashamed about it they may deny giving themselves pleasure and not tell their partner how they have achieved orgasm or show him what to do. Some women who can masturbate easily to orgasm get frustrated that their partner does not touch them in the right way and rhythm. It takes time and patience to work this out together, since men usually have very little knowledge or understanding about the clitoris. They tend to rub too hard, similar to the way they masturbate themselves, so they do need teaching. The clitoris is not a penis!

Over the years there has been enormous controversy about the clitoral versus the vaginal orgasm, usually written by men. Masters and Johnson settled the argument by finding that there is *one* orgasm – the vaginal muscles contract rhythmically whether the clitoris is stimulated directly by hand or vibrator, or indirectly during intercourse, when friction is applied by the man's pubis and the penis pulling on the lips around the clitoris as it moves in and out of the vagina. However, the intensity of the orgasm can vary depending on the level of sexual arousal (the greater the arousal, the greater the intensity) as can the perception of where the pleasure is most intense, around the clitoris or in the vagina. It may take time for the woman to perceive what is happening in the vagina. The upper two thirds of the vagina is insensitive to direct touch but has nerve endings that are sensitive to pressure. The lower third of the vagina is very sensitive to touch and, when the woman is near to orgasm, this part narrows forming the orgasmic platform which then contracts rhythmically when climax is reached.

Thus, for a woman to obtain an orgasm, it means knowing about her body and where she likes being stimulated and sharing this knowledge with her partner. It also means 'letting go', getting lost in pleasure and not being afraid to lose control. Some women who find it difficult to climax are anxious about losing control; they are afraid that it is 'not nice', that they will scream out or that they will pass urine or wind. They can let themselves get aroused and excited but at a certain point, when orgasm might occur, they switch off. This usually leaves them feeling angry and frustrated and also resentful of their partners who appear to have an orgasm so easily. There is a fear that by having an orgasm the woman will be changed into another person (a prostitute perhaps?) and that everyone will know and be disgusted. These are feelings that often go back to early childhood and negative parental attitudes about sex and losing control.

Finally, when women are involved in an unsatisfactory sexual relationship, they may need to ask themselves why that is so. Often women with poor self-esteem and self-image get caught up in masochistic relationships where they are abused or neglected. This is especially so when women have been sexually abused as children. Psycho-sexual counselling may offer the best opportunity of getting free from such relationships.

The vaginal examination

Reference has been made several times to this examination and its place in helping to treat sexual difficulties. Women may be afraid of being examined and feel exposed and vulnerable. It is sadly true that not all doctors do this examination with care and sympathy so women may have had a previous bad experience, or have heard of other women's bad experiences. The examination is important not only to familiarize the woman with her own sexual organs but also to exclude any physical abnormalities and to take a cervical smear. Many women who develop cancer of the cervix have never had a smear test because they have been too frightened to be examined. The examination may be uncomfortable if the woman is tense and tightens her vaginal muscles. If she is frightened she should tell the nurse or doctor so that extra care may be taken. If the woman prefers to see a woman doctor she will usually find that family planning and well-woman clinics are run by women doctors. There are also more women general practitioners than there used to be.

Sexual preference

What has been written so far has been about heterosexual relationships. However, there are women who are not sexually attracted to men but to other women. These are not the same as women who have *chosen* to have sexual relationships with other women as a protest against men – so-called political lesbianism. Women, like men, lie on a sexual spectrum, from those who are exclusively heterosexual all their lives in thoughts and behaviour at one end to those who are exclusively lesbian at the other. There are those women who fall somewhere in between: those who occasionally have sexual fantasies about other women but who are sexually involved with a man; and those who have sexual relations with both sexes.

For some women their attraction to their own sex may be seen as shameful or abnormal and they will need counselling to accept themselves as they are. Yet other women in this situation will not be concerned about their lesbianism and can accept it as natural and normal for themselves. Unfortunately, being a lesbian may mean facing prejudice and ridicule both from other women and men. Some men seem to find lesbianism threatening or frightening and hold the belief that once the woman has had intercourse with the 'right' man (that is, one with a large penis) she would no longer want to be a lesbian. This naive view ignores the variation found among human beings. Perhaps the most difficult areas to be faced are to do with the reactions of parents and siblings to the knowledge that their daughter or sister is a lesbian. Some parents can eventually be brought round, but others are hostile, cutting themselves off and leaving the woman feeling isolated, lonely and rejected. Married women with children who discover their attraction to other women at a late stage in their life may have to face the ordeal of proving their suitability to have custody of their children if they decide on divorcing their husbands.

Finally, perhaps, it needs to be said that lesbian relationships, like heterosexual relationships, are not immune to jealousies, rivalries, poor communication and infidelity. Since such relationships may face considerable antagonism from the rest of society, the women may become very dependent on one another so that if the relationship does break down the separation can be especially devastating and guilt-laden. There are counselling agencies specially geared to help lesbian women and lesbian mothers.

4 *Pre-menstrual Syndrome*

by Penny Chorlton

Pre-menstrual syndrome is the term that may be more familiar to most people as PMT – pre-menstrual tension. The umbrella-word 'syndrome' is more accurate these days because doctors now realize there are anything up to 150 physical and mental symptoms associated with the changes that lead up to menstruation, not simply feelings of *tension*. It is because there are so many different symptoms that many doctors still do not know how to recognize PMS for what it is and thousands of women still go undiagnosed, untreated or worse – mistreated.

One of the most tragic aspects of this mistreatment is the fact that thousands of women have been put on tranquillizers by their doctors, suffering the permanent side-effects these bring, simply for the temporary relief they need during the two weeks before menstruation. The unfortunate result of this is that many women who were genuinely suffering from PMS either had to submit to inappropriate treatment or struggle along – often leading a 'Jekyll and Hyde' existence and having to cope alone with their problems.

Fortunately, these days we have learned rather more about PMS and there are various forms of treatment available. However, research still continues for more clues and better treatments for the hundreds of thousands of women who are sufferers.

It is important to realize that there is no 'cure' for PMS, but rather a series of possible treatments, which either treat individual symptoms, e.g. diuretics for the feelings of bloatedness associated with water retention, or which help alleviate the hormonal imbalances that lead to a variety of unpleasant symptoms. Since every woman is different, it will be a matter of self-exploration and trial and error before a woman hits upon the right course of action or treatment for her. Hormone therapy will work for one woman, while another, even with identical symptoms, will find such help useless, but total relief with taking B6.

Who suffers PMS?

Probably everyone suffers PMS – not directly of course. In women, its incidence is actually unknown, but a consensus suggests that perhaps *half* of all women suffer from it, with between ten and twenty per cent suffering severely enough to seek help from their doctors. For this minority, PMS can become like a canker which destroys their social and family lives and their careers and which, when untreated or undiagnosed, can render years of otherwise productive and fulfilling life an absolute hell for the women themselves and the people close to them.

PMS can strike suddenly – women who have never suffered any problems at all can suddenly find themselves feeling unwell. Indeed up to eighty per cent of women will suffer from PMS at some time in their lives. As with dysmenorrhoea (period pains) there is a tendency to assume that never having suffered is synonymous with being totally exempt from it. It frequently strikes women who have never had any problems with menstruation and it is often the last thing they think it can be. Several studies show that it is often the partners of such women who suggest PMS as a possible cause of their erratic behaviour, rather than doctors or even the women themselves.

PMS is more common in women in their thirties and often comes on after a pregnancy, when it can be confused by the woman herself and doctors as postnatal depression. It is more common among married women than single women and it has even been cited in divorce cases. Lack of outdoor physical exercise makes PMS worse, as do poor eating habits. It tends to run in families, but not necessarily, and it is more noticeable and harder to bear during periods of stress.

It affects *everyone,* because study after study has shown how the husbands, children, relatives, friends and business colleagues 'suffer' at second-hand the symptoms of a woman who is 'not her usual self' because of PMS. Husbands or sexual partners usually bear the brunt of PMS-inspired hostility and aggression.

What is PMS?

PMS is a complex set of up to 150 different symptoms, both psychological and physical, which occur *only* in the two weeks prior to menstruation and which are relieved completely within forty-eight hours of starting to bleed. It is the actual timing of these symptoms that can determine whether a migraine, irritability or depression is because of PMS or something else altogether. Diagnosis, therefore, is most accurately performed by the woman herself, who should keep a menstrual diary and note down any symptoms to see whether they form a cyclical pattern over a period of months.

Since these symptoms tend to come together in groups, some researchers have started to label 'sub-groups' of PMS and, in years to come, it may well be that women will be able to identify what variety of PMS they suffer.

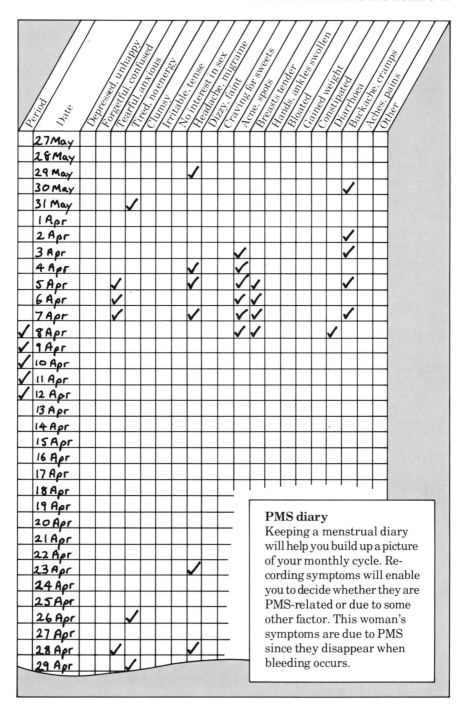

PMS diary
Keeping a menstrual diary will help you build up a picture of your monthly cycle. Recording symptoms will enable you to decide whether they are PMS-related or due to some other factor. This woman's symptoms are due to PMS since they disappear when bleeding occurs.

PMS: The most common physical and psychological symptoms

- Anxiety, tension and worry, even paranoia
- Irritability, hostility and anger – often uncontrollable
- Depression, tiredness, unhappiness, weepiness, feelings of insecurity
- Lethargy, loss of energy
- Food cravings – especially chocolate and sweet things
- Swollen, tender, painful breasts

- Swollen ankles
- Weight gain, bloatedness, swelling of the abdomen, cramps
- Headaches, migraines
- Skin breaking out, acne
- Painful joints or aching back
- Clumsiness, breaking things, memory lapses, confusion
- Constipation or diarrhoea
- Loss of interest in sex
- In extreme cases, suicidal feelings

Because of the sheer range of symptoms, it is perhaps understandable why some doctors overlook PMS. Often, such doctors only treat specific symptoms, like prescribing diuretics for water retention, or tranquillizers for depression, or painkillers for migraine attacks.

What causes PMS?

The hormonal theory

PMS is caused by a hormonal imbalance in the body which takes place regularly each month as the body adjusts itself in readiness for the shedding of the lining of the womb – menstruation. Unlike men, whose hormone levels do not fluctuate, women's hormones are changing all the time, causing certain alterations in the body throughout their reproductive lives. These start at puberty and end with the menopause.

Dr Katharina Dalton, one of the world's experts on PMS, is convinced that it is caused by the drop in progesterone levels that takes place in the last two weeks of a woman's cycle. Her solution is to replace the progesterone during this phase by prescribing the synthetic hormone, and so reduce the symptoms. This works for some women, but other experts say that something like sixty to seventy per cent of PMS sufferers have normal progesterone levels, so this cannot be the right answer for them. This is something that can only be measured by a laboratory test.

Two women who had been helped by this form of treatment were so impressed that in 1983 they set up a self-help organization called the National Association for Pre-menstrual Syndrome (see Useful Addresses), which has been organizing and setting up support groups around Britain. These can be very therapeutic for PMS sufferers and provide a useful forum for exchanging ideas on how to cope with and reduce the symptoms associated with severe PMS.

The nutrition theory

In recent years, a number of experts have developed the theory that PMS is a condition caused by nutritional imbalance. They have shown that women who have a deficiency in essential fatty acids suffer the same sorts of symptoms as found in PMS. Essential fatty acids (EFAs) are otherwise known as polyunsaturates and they cannot be made in the body, but have to come from food. If you are eating the wrong food, the argument goes, you are depriving your body of what it needs to function properly and you may suffer the symptoms associated with PMS.

Unlike all other approaches to PMS, the nutritional one gives back to women control over their bodies and lifestyles. They can alter what they eat and discover whether there is any improvement. Oil of evening primrose is an EFA, and many – though not all – women find it very useful as a food supplement that allows them to eat as before while correcting any imbalance and thus minimizing or 'curing' PMS. A substantial number of studies in Britain and in other parts of the world have established that for the majority of women suffering from PMS oil of evening primrose is the answer to their problem.

Evening primrose oil is made and sold by several companies and may be expensive as a permanent form of treatment but women who would otherwise be in severe discomfort may view it as a worthwhile expense. Unfortunately, neither it, nor anything similar, can be prescribed on the NHS. Few people would want to elevate PMS into an 'illness' so treatments for it, especially self-administered ones, will almost certainly always remain something women will have to pay for themselves. However, if women attempt to improve their dietary habits they might no longer need to take supplements like evening primrose oil.

The nutritional approach to PMS is now very popular and various groups exist to help women to help themselves. Too much calcium, for example, as found in dairy products, requires more supplies of magnesium for the body to work efficiently and, unless these can be found, or supplied, PMS symptoms may occur. Similarly, tea, coffee and chocolate all use up the B vitamins in our bodies and, if you consume a lot of these, you will need to supplement your B6 intake.

Other contributing factors

Smoking

Women now smoke far more than they used to and the rate at which they take up and increase the amount of cigarettes they smoke is becoming greater. Smoking not only depletes some of the body's natural defences, in particular vitamin C, but also makes some PMS symptoms worse – in particular, breast tenderness.

Alcohol

Again, women are drinking far more alcohol than they used to and, overall, consumption in Britain has increased by nearly 100 per cent since the Second World War. Alcohol depletes most of the minerals and vitamins needed by the body and it stimulates urine production and so washes out all nutrients that dissolve in water.

Sugar

The British are the largest consumers of sugar in the world. They eat astonishing amounts of sugar in processed foods and as sweets and chocolates. Sugar blocks the body's absorption of vitamins and minerals and causes 'stop/start' energy levels.

Coffee and tea

Consumption has again risen dramatically since the Second World War because coffee has become relatively cheap. The caffeine in coffee and, to a lesser extent, in tea and fizzy drinks is addictive. Yet caffeine aggravates breast tenderness and the tension associated with PMS, as well as making people more anxious (mistaken for alertness) and insomniac. It has been estimated that each cup of coffee contains 100–150mg of caffeine – the same amount as in four aspirin tablets. Coffee devotees should switch to decaffeinated brews and tea drinkers should try and have just two cups a day – and then drink it weak with lemon, not milk.

The Pill

The Pill, which is taken by around three million women in Britain, can interfere with the vitamins and minerals in the body, especially the B vitamins and vitamin C. Studies have shown that around half the women taking the Pill are deficient in vitamin B6 and this can make pill-takers particularly prone to depression and mood swings.

PMS and work

Oddly enough, studies have shown that women doctors are just as likely as their male colleagues, if not more so, to be dismissive of PMS. This might be because, as career-women, they have had to overcome the problems associated with being a woman, and they do not like to indulge other women in feelings of not being able to cope without medical help.

Certainly many career-minded women are very reluctant to acknowledge the existence of PMS, or that they might succumb to its negative influences, for fear that this may set them in a bad light against their male colleagues. Studies have shown that women will pretend to their employers that they have suffered anything other than PMS if they have been forced to take time off, because of fears that it could handicap their career if it were realized that they suffer from a condition that affects them for two out of four weeks a month.

Dr Caroline Shreeve, who has specialized in PMS, recommends that women who suffer from it should arm themselves with a PMS kit to keep near them at work. Crazy as it sounds, this includes a pair of sunglasses, both to protect them against the aggravation of fluorescent lighting – which can be a nightmare for those prone to headaches and migraines, with or without PMS – and also to hide behind in case of tears or anger associated with PMS outbursts. Her suggested PMS kit also includes sweets and tempting snacks to stave off hunger pains, aspirin or painkillers, and worry-beads for dispelling tension. For women who suffer irritation, she suggests they clear their desks and try to get ahead of their work, if possible, in the run-up to the PMS phase and, once PMS has started, they use the lunch hours to relax or, better still, take some physical exercise. In some cases, however, the best solution may be simply to stay at home and rest in bed.

Many women resent PMS being described as an 'illness' or 'disease' simply because it is so commonplace and it is absurd to think that half the world's female population are 'suffering' from such a widespread but ordinary condition. For the sake of convenience only, in this chapter PMS may be referred to as an illness, but only where the term seems appropriate.

'Jekyll and Hyde'

For most women, the single most distressing and sometimes frightening symptom of PMS is loss of control of their feelings. Time and again, women, especially mothers, report their horror at the way another personality seems to take them over during the worst phase of the run-up to their period. Otherwise gentle wives and mothers have become violent, and have even been known to attack children and spouses with weapons.

Occasionally, women have committed crimes under the influence of PMS and it has been recognized as a mitigating factor, but not a legal defence. The issue has been controversial; (male) judges have so far been convinced that if PMS were allowed as a blanket defence, like insanity, women could be theoretically above the law and free to break it, as long as their crimes were unpremeditated and carried out in the PMS phase.

Dr Dalton, who runs a private practice in London's Harley Street, studied the crimes carried out by a group of women prisoners in Holloway and found that nearly fifty per cent had committed their crimes within four days of menstruating, that is, in the PMS phase of their cycles. Dr Dalton will appear for defendants but only after she has interviewed them, spoken to their doctors and taken a full medical history to establish a history of severe PMS. It is not a defence any woman can look to but, for a few women, PMS has been shown to lead to uncharacteristic bouts of spontaneous and uncontrollable criminal activity – shoplifting being one of the most common crimes. Dr Dalton turns down many lawyers and women who ask her to be an expert witness as she is most concerned that non-PMS sufferers do not jump on the bandwagon.

Where genuine PMS has been proved to the courts, some women have been discharged and not held responsible for their actions, while others have received lighter sentences than would have been appropriate had they not been under the influence of PMS.

Getting help

Unfortunately, there are very few PMS clinics run by the National Health Service, but the Family Planning Association has one or two, and the addresses at the end of the book may help you get in touch with a clinic, or even an on-going research group.

Hormonal treatments

Natural progesterone
Dr Dalton's method, this is widely available through GPs. It cannot be taken orally, so has to be prescribed by a doctor as either injections or suppositories. Doses of 200–400mg are taken, one to two weeks before menstruation.

Dydrogesterone
This is a compound closely related to natural progesterone. It can be taken in tablet form, again for only part of the cycle.

The Pill
Because the contraceptive pill contains synthetic hormones that replace the activity of the natural hormones in the body, the Pill usually depresses normal hormonal functions and, with them, the symptoms of PMS. Unfortunately, women who have PMS often have a noticeable difficulty in tolerating the Pill.

Oestrogen 'Oestradiol' implants
These are available for a small number of women for whom no other treatment works. Many doctors think the treatment, which involves implanting the hormone oestrogen, is too drastic. A recent research programme at King's College Hospital in London has shown that the treatment does work, but that it is probably only suitable for severe cases.

Change your eating habits to avoid PMS

Cut down on these
- Dairy products, to a maximum of two servings a day (too much calcium interferes with magnesium absorption)
- Tobacco/nicotine
- Refined sugar
- Tea and coffee, and leave out sugar or sweeteners if possible (herb teas can act as a diuretic which is useful for bloated feelings)
- Salt, and salty and spicy foods, because these aggravate water retention
- Red meat, to three ounces a day maximum
- Fats, especially cooked and saturated e.g. butter, lard, dripping

● Eat more of these
- Green leafy vegetables, legumes (peas and beans), and potatoes
- Cereals, whole-grains (corn, millet, rye, wheat, brown rice, barley etc.)
- Nuts, most varieties
- Polyunsaturates, e.g. safflower oil, evening primrose oil etc.
- Fish, especially oily fish like herring, sardines etc.
- Chicken and other poultry
- Fruit, fresh and dried
- Low-fat cheese and yoghurt, and skimmed milk

The good thing about eating a diet that reduces PMS symptoms is that it is also a healthy diet generally. For example, most of us eat twenty times more salt per day than we actually need and we nearly all eat far too many saturated fats, too much red meat, and too much sugar. All the foods PMS sufferers are recommended to reduce or increase will also mean they are at less risk of heart disease and all the other illnesses associated with a poor diet. As an even more gratifying bonus, a diet that is healthy for PMS is also a slimming one!

Moira Carpenter, founder of the Pre-menstrual Tension Centre, and author of *Curing PMT the Drug-free Way,* suggests that everyone should eat at least half their food *raw*. She also suggests that people who want to eat the right things but who do not have the time to shop and plan menus should add the following to their food to make up for any deficiencies:

- Brewer's yeast, contains B vitamins and is high in protein
- Linseed, rich in linoleic acid (an essential fatty acid)
- Wheatgerm, a natural source of B vitamins, manganese, essential fatty acids and vitamin E
- Kelp, a powdered form of seaweed containing many minerals, including iodine
- Lecithin, a prime source of polyunsaturated fatty acids

All these can be purchased from health food shops and good pharmacies. Once opened, vitamins only stay active for about a year so it is best for women to buy as they go and not stock up erratically. They are not cheap, but they enable women to carry on with their normal lifestyle while reducing the severity of PMS symptoms and, at the same time, improving their general level of health.

PMS craving for sweet things

Many women suffering from PMS crave sweet things in the run-up to their period, usually in the form of chocolate. Stress often aggravates this craving. A couple of hours after eating the chocolate, however, they start to feel low, fatigued, and they may get a headache or 'the shakes'. This is because when a person feels stressed, and eats refined sugar to give themselves a 'lift', the stress creates a deficiency in dopamine, the enzyme in the brain that controls sedation. The highly refined sugar forces tryptophan into the brain cells where it is converted into serotonin, and too much of this causes nervous tension, palpitations, drowsiness and an inability to concentrate.

Why chocolate? Well, chocolate is rich in magnesium and without this ingredient, the body cannot break down sugar and derive energy from it. So the craving for chocolate may really be a sign of magnesium deficiency and it could be a natural instinct that is actually right, being rewarded with something that is quite wrong. A better response would be to eat something else rich in magnesium, which does not produce unpleasant effects, such as whole-grains, green leafy vegetables and legumes. Unfortunately, when you are out and about or at work, or in a hurry, it is much easier to grab a bar of chocolate than cook up a plate of runner beans.

Vitamins and minerals: how much do you need?

Vitamin B6

There is no official DHSS recommended daily allowance (RDA) for vitamin B6 (which can be regarded as the anti-depression vitamin) but Americans are advised to take 2mg a day. Statistics from the Ministry of Agriculture, Fisheries and Food show that the modern British diet contains less B6 than the average diet of thirty years ago. Other studies have shown that women tend to have too few B6 reserves, which normally does not matter but which may become an issue when vulnerable to PMS.

Women who take the Pill have been shown to have serious vitamin B6 deficiencies. It can also interfere with and deplete other vitamins and minerals in the body. Therefore women who take the Pill and who suffer from PMS should try taking B6 first to see if it helps.

Symptoms of vitamin B6 deficiency include insomnia, irritability, depression, anxiety, and other psychological changes. Other symptoms can be dandruff, acne and rough red pimples on the upper arms and thighs, and redness and greasiness on the sides of the nose.

Iron

Women, because they lose blood each month in menstruation, need the mineral iron more than men. Yet several studies have shown that the majority of women in Britain have levels of iron below the officially recommended 12mg a day (which is itself way below what other countries estimate is the ideal). People who eat a lot of processed foods do not usually get enough iron, especially women and children.

The symptoms of iron deficiency are lethargy, restlessness, muscle weakness, palpitations, indigestion and a reduced resistance to infections. Physical signs include brittle nails, sore tongue and cracking at the corners of the mouth.

Zinc

Zinc deficiency is now very common among the British population – partly because it is absent from processed foods. Again, Britain has no official recommendation, although the World Health Organization suggests people should have 11mg a day. American experts recommend higher levels and as much as 20mg a day for pregnant women. Again, studies show that British women do not have sufficient supplies of zinc.

The symptoms of zinc deficiency include depression, irritability, impaired sleep and loss of appetite. Physical signs include acne, white spots on the nails, hair loss and facial dermatitis.

Magnesium

There is no official recommended daily intake of magnesium in Britain, yet several studies have shown that British people have way below the amounts recommended in America because they eat less of the grain and green leafy vegetables in which magnesium is found in large quantities.

The symptoms of magnesium deficiency are loss of appetite, apathy, nausea, and muscular cramps.

Supplements for dietary deficiencies

B6 (Pyridoxine)

The normal dose is 40mg twice daily, for three days before the expected start of any PMS symptoms. This is enough for most women but, in more severe cases, it can be increased to 75mg a dose, making 150mg a day in total. B6 is taken for about five days in all, and stopped halfway through the period. Extra top-up doses can be taken but it is important that women should not overdose themselves as too much vitamin B6 can be dangerous.

The danger lies in the fact that B6 is something women can buy over the counter, without a prescription, and it is cheap. The natural tendency is to take more of something if it appears not to be helping and, since B6 is preventative rather than curative, women often give themselves megadoses of 500mg a day just to make sure. Doing this over a long period is harmful and should not be done by anyone unless they are under the medical supervision of a doctor who knows what he or she is doing. Any doses in excess of 100mg a day can cause gastric problems and there are more serious side-effects still if a woman is megadosing and pregnant – some women claim that their babies have been damaged or born dead because they had been taking excessive quantities of B6 during pregnancy. Magnesium, incidentally, can reduce the side-effects of B6.

Magnesium and zinc

Because of the dietary deficiencies commonly found in women suffering from PMS, some experts believe that supplements of magnesium or zinc can help. A better source would be from natural foods but supplies are not always readily available, and so there is a case to be made for such supplements.

Other special PMS supplements

A variety of specially-formulated PMS vitamin and food supplements are now available through health food shops and by mail order. Check carefully what is in them and, if you are at all worried, take a doctor's advice. Doctors warn women to be careful what they take in the way of supplements, because too many can cause further or different imbalances in the body. A woman should avoid taking anything like this when pregnant.

Correcting faulty nutrition is a first basic step, and should be considered by women as the sensible way to deal with the unpleasant symptoms of PMS. And only if this fails should they consider resorting to drugs, not the other way round. After all, drugs do not get rid of the symptoms permanently, they merely suppress them.However, changing your eating habits cannot be done overnight – new, better habits have to be learned, which takes time and effort. Do not be surprised if it takes months to diagnose individual deficiencies and then get used to new ways of eating, but your efforts are likely to be rewarded in more ways than one.

5 *Birth Control*

by Toni Belfield

The method of contraception a woman chooses and how she uses it is to a large extent tied up with how she feels about herself sexually, her degree of comfortableness with her body, and her knowledge of how her body works. This choice may also be dependent on her age, culture, lifestyle, needs and general health. Any decisions she makes may or may not involve her sexual partner. It is also important to realize that a woman may change methods throughout her reproductive life as her needs alter.

However, most importantly, the safety and effectiveness of the various methods will affect the decision a woman reaches. Research to date has been slow to investigate ways of making safe methods more effective, or effective methods safer. Therefore, it seems for the present that there will always be a 'trade off' between effectiveness and risk. This is well illustrated by the millions of women who choose the Pill rather than the cap or natural family planning methods.

There is a lot of confusion about what is meant by effectiveness. In this chapter a range of effectiveness is given for each method of contraception. The higher figure represents the efficiency of the method when used correctly and consistently. This drops to the lower figure where the method is used incorrectly or sporadically.

What is family planning?

Family planning is not about not having babies, it is, as it says, about planning families. This not only means limiting the number of children and timing their arrival (birth control), but also involves *planning* a baby: timing conception, pregnancy care, as well as help and advice for fertility problems. Today, family planning also embraces reproductive health care; this includes well-woman/man checks and providing help for sexual, menopausal and PMS problems.

Where to go for birth control

In the UK, there is a choice of where to go. Birth control advice and supplies are free on the NHS from family planning clinics or most GPs.

Family planning clinics provide a comprehensive range of methods for men and women, as well as reproductive health care services. Some clinics run special 'youth' advisory sessions. Addresses of clinics are listed under 'family planning' in the telephone directory, or you can telephone the Family Planning Information Service (FPIS) for details (see Useful Addresses). On your first visit, your name and address, as well as your GP's, will be taken. (You can tell the clinic not to give information to your GP without your consent.) The doctor or nurse will discuss the various methods of contraception and answer any queries or concerns you may have. Relevant details about your own and your family's medical history will be noted, and your blood-pressure and weight will be checked. You will also be offered an internal examination, cervical smear test and breast check. These can be done on another visit if preferred.

Most GPs offer family planning. If your GP doesn't, or if you do not want to go to him/her, you can go to another GP who provides this service. Alternatively, you can go to a family planning clinic. Not all GPs fit caps, diaphragms and IUDs, or offer natural family planning. GPs at present cannot prescribe the sheath or sponge.

If you cannot travel to a clinic or to your GP for any reason, a domiciliary visit can be arranged. This is when the doctor or nurse comes to you.

Other services

There are various non-NHS services in the UK (some are private, some are charitable) offering a wide range of reproductive health care services and birth control. These include, Marie Stopes House and the British Pregnancy Advisory Service (BPAS) (see Useful Addresses).

Oral contraception – the Pill

Oral contraception is always an emotive subject, as the Pill is a drug taken by healthy women, often for long periods of time. However, few things have been subjected to such scrutiny by research, the media and, most importantly, by women themselves.

Although we always refer to 'the Pill', there are two kinds: the combined pill, which contains two synthetic hormones, oestrogen and progestogen; and the mini-pill, which contains only progestogen. Different brands have different formulations and amounts of oestrogen and progestogen, so if one pill does not suit you, it's worth trying another.

The combined pill

Combined pills are available in two main forms and are ninety-three up to ninety-nine per cent effective: the standard monophasic (fixed dose) pill, where each pill is identical; phasic pills (Biphasic and Triphasic), where the ratio of oestrogen to progestogen varies through the cycle, twice for biphasics and three times in triphasics.

The Pill works by

- preventing ovulation
- altering the lining of the womb to prevent implantation
- altering the cervical mucus so that it cannot be penetrated by sperm
- changing the movement of the fallopian tubes, which affects the passage of the egg.

Using the Pill

Combined pills come in bubble packs of 21, 22 or 28 pills. One pill is taken each day for twenty-one or twenty-two days, followed by a break of seven or six days respectively. During the pill-free days you will have a 'withdrawal' bleed, which is lighter and shorter than a normal period. If the Pill has been taken correctly, you are protected against pregnancy during these days. At the end of the pill-free break you should begin the next packet. The 28-day pill packs (called EveryDay or ED pills) contain seven dummy pills and bleeding occurs during this time. All types of pill should be started on the *first* day of a period. No extra contraceptive precautions are needed (except for ED pills, where you are not covered for the first *seven* days). If you start at any other time you are not contraceptively covered for the first *seven* days of the pack. Phasic and ED pills must be taken in the right order and are colour coded to make this easier.

Forgetting one or more pills, having sickness or severe diarrhoea, or taking certain drugs (some antibiotics, anticonvulsants, and drugs for tuberculosis) can all reduce the Pill's effectiveness and you will not be contraceptively covered. Where pills are missed you should continue to take them as normal, and use another method such as the sheath for the next seven days. If sickness or diarrhoea occur, or you are taking medication, continue to take your pills and use extra precautions over this time and for the following seven days. If these days of extra precautions run into the pill-free week, you should run on *immediately* with the next packet, that is, leave no break between packs. (If using ED pills, miss out the dummy pills.) To change from one type of pill to another, just start a new packet immediately, that is, leave no break between packs.

Advantages

The Pill is the most reliable, convenient, reversible, non-intercourse-related method and can be used until the age of forty-five if you have no problems or do not smoke. In addition to providing relief from painful periods and reducing PMT, the Pill can also guard against risk from the following:

- Anaemia
- Benign breast disease

- Ovarian and endometrial cancer
- Uterine fibroids
- Rheumatoid arthritis
- Endometriosis
- Thyroid disease
- Pelvic inflammatory disease (PID)
- Ovarian cysts

Disadvantages

- The most important risk is thrombosis (formation of a blood clot in the arteries or veins). While uncommon, the risks are greater in older women and women who smoke.
- High blood-pressure. Most women show an increase in blood-pressure when on the Pill but in some cases it rises too much. This is why your blood-pressure is checked each time you visit your clinic or GP.
- Depression. This is not common with modern pills, but it seems to be related to body chemistry. Some women on the Pill have lowered Pyridoxine (vitamin B6) levels and supplements may help.
- Lowered sex drive. This is often related to depression. Sometimes the Pill is blamed when another problem is the cause. If pill-related, a change of brand may help.
- Headaches. These sometimes occur during the first few packs and should not last. Any unusual headache should be investigated.
- Nausea. Feelings of sickness may occur when first using the Pill, and, again, should not last.
- Weight gain. Although uncommon with today's pills, this can occur during the first couple of packs, but should not last. Some pills may increase appetite.
- Breast tenderness. This can be helped by changing brands.
- Vaginal discharge. This seems to be common in some women and, although harmless, can be annoying.
- Future fertility. The return of natural periods after using the Pill can take some time for a few women, and is worse for those who come off the Pill in their thirties.
- Skin changes. Some women get a brown discoloration on their faces known as the 'pregnancy mask' since it can occur during pregnancy.

The following women should not use the combined pill

- Those who think they are pregnant
- Women who have abnormal vaginal bleeding
- Those who have had a previous thrombosis
- Women with heart disease or a family history of this at an early age
- Those suffering from liver disease
- Migraine sufferers taking ergotamine-containing drugs

How safe is the Pill?

Do press reports make you panic? If they do, you are in the company of hundreds of other women. All drugs have side-effects and the Pill is no exception. In order to minimize these, the hormonal content of the Pill has been steadily reduced over the years, so that today's pills contain the lowest doses to give effective contraceptive cover and good cycle control. The problem now, according to one researcher, is that women metabolize the Pill in different ways, so what is safe and effective for one, may not be for another.

The Pill and cancer

Since the connection between cancer and taking the Pill is still uncertain, the debate is obviously of concern to pill-users. The problem is that no one has the answers yet. The literature is copious, complex and contradictory, often looking at older women using higher-dose pills. Recent studies have suggested a link between using the Pill and the development of breast, cervical and liver cancer (the latter is rare). This risk may be higher in women who use the Pill for a long time. A number of studies are now under way to provide a clearer picture.

See your doctor if you develop any of the following symptoms

- Painful swelling in the calf
- Pain in the chest or stomach
- A bad fainting attack or collapse
- Unusual headache, or disturbance of speech or eyesight
- Any numbness or weakness
- Jaundice (yellowing of skin or eyes)

The mini-pill

The mini-pill is an alternative if you want to use hormonal contraception but are unable to use oestrogen-containing pills, and is ninety-six up to ninety-nine per cent effective. The pills are taken continuously, without a break, and work by

- sometimes preventing ovulation
- altering the cervical mucus so that it cannot be penetrated by sperm
- altering the lining of the womb to prevent implantation
- changing the movement of the fallopian tubes, which affects the passage of the egg.

Using the mini-pill

Since it works mainly by thickening cervical mucus, the mini-pill must be taken strictly on time each day. Contraceptive protection is lost if the pill is taken more than three hours late. It is started on the first day of a period, without the need for extra precautions. Sickness, severe diarrhoea, and certain drugs (anticonvulsants and drugs for tuberculosis) can interfere with the mini-pill, but antibiotics can be safely taken at the same time. If one or more pills are missed, extra precautions should be used for the following forty-eight hours, as well as continuing to take the pills. If sickness or diarrhoea occurs, extra precautions should be used over the time of illness and for the next forty-eight hours, as well as continuing to take the pills.

Advantages
- It is easy to take.
- It is not related to intercourse.
- It can be taken until the menopause.
- There is no delay in the return of fertility.
- There are fewer adverse effects compared to the combined pill.

Disadvantages
- The mini-pill can cause menstrual irregularity, spotting or no periods at all.
- There is an increased risk of ectopic pregnancy (pregnancy occurring outside the womb, usually in a fallopian tube).
- Some women develop ovarian cysts.
- There is a higher pregnancy rate than with the combined pill, although it is very effective in women over forty.
- You must be rigorous about taking it on time.

The injectables

There are two injectable contraceptives available in the UK: Depo-Provera, which is given every twelve weeks; and Noristerat, which is given every eight weeks (short-term use only). Both contain synthetic progestogens, but work in a similar way to the combined pill and are nearly one hundred per cent effective.

Opinions differ about the value of this method of contraception. Depo-Provera has been subject to considerable controversy about its use. Injectables should only be given to women who cannot use any other method. Always obtain full information about the risks and benefits *before* the injection is given.

Advantages
- It is the most effective contraceptive method.
- One injection lasts a long time.
- It is not related to intercourse.

Disadvantages
- Once you have had the injection the hormone cannot be removed.
- Menstrual disturbances can occur and bleeding may be frequent, irregular, heavy, continuous or absent.
- There is often a long delay in the return of fertility.
- Some women suffer with depression, back pain and headaches.

Implants and vaginal contraceptive rings

These are the newest hormonal contraceptives about to arrive in the UK. The advantage of these 'slow releasing' methods is that, unlike the Pill which is swallowed and digested, they bypass the gut. Consequently they produce fewer side-effects and therefore have fewer risks.

Implants: Consisting of a number of tiny plastic rods inserted just under the skin of the upper arm, they release progestogen directly into the blood-stream and are over ninety-nine per cent effective. They have a similar action to the mini-pill and in some cases prevent ovulation. They last for up to five years.

The ring: This is a 6cm plastic ring that fits in the vagina. It slowly releases progestogen and is about ninety-eight per cent effective. It is effective for up to three months and works like the mini-pill.

Intrauterine contraceptive devices (IUDs)

Contraception that can be left in place for a number of years, which does not interfere with sexual intercourse and which is effective (ninety-six to ninety-nine per cent), must be seen as fairly ideal. These are the advantages of the IUD. The IUD is a small device about one and a quarter inches long, which is fitted into the womb and left in for a number of years according to type. There are a number of varieties, but they can be broadly divided into three groups: plastic IUDs (inert – they have no active substance in them); plastic and copper, or copper and silver (bioactive); and hormonal-releasing (medicated, not generally available, new ones being researched). Plastic IUDs, the Lippes Loop and Saf-t-Coil (the latter two are no longer available) were known as 'first generation' IUDs. These have now been replaced by smaller, modern bioactive devices which come in various shapes and sizes.

All IUDs have a 'tail' of one or two coloured threads that hang just into the vagina from the cervix. This enables you to check that the IUD is in place, and helps removal. Even today, it is uncertain how IUDs work but they probably prevent pregnancy in a number of ways. They work mainly by altering the lining of the womb in such a way that a fertilized egg cannot implant. Copper also acts as a spermicide. Hormonal IUDs actually stop ovulation as well as preventing implantation.

Having an IUD fitted
It has to be fitted and removed by a trained family planning doctor. This is usually done during menstruation; the cervix is softer at this time which makes insertion easier. There is also no risk of you already being pregnant. Once the doctor has checked the size and shape of your womb, and that there are no problems, the IUD is fitted. It is placed in the womb using a special introducer (a thin hollow tube holding the IUD). This takes only about three or four minutes. Most women find this an uncomfortable procedure: cramping (like bad period pains) and bleeding can occur.

Ask to see your IUD before it is fitted. This gives you an opportunity to know which kind you have and how small it actually is. Check with the doctor when it needs to be replaced. You will be taught how to feel the strings: these should be checked regularly in the first few months and then once a month after each period. The risk of the IUD falling out (expulsion) is greatest just after it has been fitted. If you feel anything other than the strings, always get it checked. The strings should not be felt during lovemaking, but if they have been cut too short they may stub the penis, which will be very uncomfortable for your partner. Usually they are left long enough to tuck themselves around the cervix out of harm's way. The golden rule about the IUD is that if you are worried about anything, always go back and get it checked.

Advantages

- Contraception becomes effective immediately.
- IUDs are very suitable for women who are spacing out their pregnancies, or who have completed their families, or for older women.
- They are not related to intercourse.
- There is nothing to remember to do (except to check the strings).
- IUDs are very effective.
- They only require a yearly checkup.
- They can be left in place for a long time.

Disadvantages

- IUDs can cause menstrual irregularity, or more frequent or heavier bleeding (especially when first fitted).
- A watery discharge is common.
- There is a risk of developing pelvic infection.
- Ectopic pregnancy becomes a higher risk.
- There is a greater chance of miscarriage if pregnancy occurs with an IUD in place. (But there is no evidence of any handicap or abnormality in the baby if the pregnancy goes to term.)
- There is a rare chance of perforation of the womb (about one in 1,000).

Pelvic infection

This is the most serious risk associated with using an IUD. Any unusual bleeding, pain, unusual vaginal discharge, pelvic tenderness or raised temperature occurring with an IUD in place could signal a pelvic infection. Any symptom like this should be investigated *immediately*. Pelvic infection can be easily treated with antibiotics. If left untreated, it can lead to damaged fallopian tubes and infertility. IUDs do not cause, but facilitate, the transmission of infection or a sexually transmitted disease. Infection can occur during fitting, but this is unusual. Infection occurs mainly in young women who have many sexual partners (or more than just one), or whose partner has other sexual contacts. For this reason it is not a recommended method for women who are not in a steady relationship and/or who have not had children. An IUD (Dalkon Shield) fitted in the early

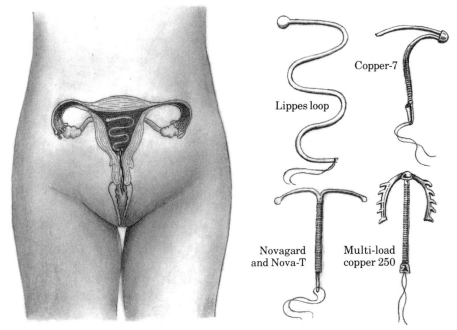

Copper-7

Lippes loop

Novagard
and Nova-T

Multi-load
copper 250

Diagram showing an IUD in place, with the threads hanging into the vagina from the cervix. The four IUDs on the right are shown actual size.

1970s was specifically linked with infection and even caused some deaths. Although it was withdrawn from the market, some women may still be wearing this type today. If you had an IUD fitted at about his time, and you are not sure what it is and it is still in place, go and have it checked.

The following women should not use the IUD
- Those with past or present pelvic inflammatory disease or infection
- Young women who have not had children, or who are not in a steady relationship
- Women who think they are pregnant
- Those with undiagnosed vaginal bleeding
- Sufferers of cancer of the cervix or womb
- Women with previous ectopic pregnancy
- Those with large fibroids, or an abnormally shaped womb

The barrier methods

These methods physically prevent the passage of sperm to the egg. They include some of the oldest attempts at fertility control which have been used for centuries. Today's barrier methods – the sheath, diaphragm and cap – declined in use with the introduction of the Pill and IUD, as they were considered old-fashioned and intrusive. However, as more is learned about the potential hazards of newer methods, and with more questions being asked about our health and lifestyles, there is now a revival of interest in 'the barriers'.

The cap and diaphragm

One myth worth dispelling is the idea that there is just one cap for women! We talk loosely about using the 'cap', but in fact there are two main types:

Diaphragms: These are shallow domes of rubber mounted on a flexible metal ring which are inserted into the vagina and cover the cervix and surrounding area. There are three kinds: flat spring, coil spring and arcing spring diaphragms. They come in various sizes.

Caps: These are smaller, come in various sizes and fit directly over the cervix by suction. There are three kinds: cervical, vault and vimule caps.

Both diaphragms and caps must be used with spermicides to be effective (eighty-five to ninety-eight per cent). Because diaphragms and caps are used in a similar way, they will be referred to simply as 'caps'.

Obtaining the cap
The correct size and type for you can only be determined by a vaginal examination. This is carried out by a family planning doctor or nurse. You will be taught how to insert, remove and care for your cap. Because it takes some time to use a cap correctly, you will probably be given a 'practice' cap. This should not be used as a contraceptive. Once you feel happy about using it, you will be given a new one for contraception. You should have your cap checked every six to twelve months, and after childbirth, a miscarriage, abortion or after a weight gain or loss of 3kg.

Using the cap
Myth has it that the cap has to be inserted in the last few seconds before intercourse. In fact, you can put your cap in at any time that suits you. It can be left in all day without causing any harm. (It should, however, be removed and cleaned after twenty-four hours.) The only rule is that if it goes in more than three hours before intercourse, or if you have intercourse more than once, you should top up the spermicide. But don't let putting in the cap be a problem – insert it as part of loveplay or get your partner to insert it for you!

Advantages
- Caps are very effective with careful use.
- There are no adverse health risks.
- They may protect against some STDs, pelvic inflammatory disease and cancer of the cervix.

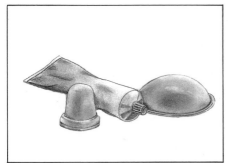

Vimule cap, spermicide and diaphragm

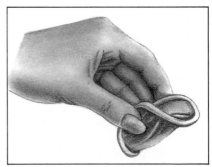

Holding the cap

Disadvantages
- You need to plan ahead so that the cap is in place before intercourse.
- There is a need for regular checks.
- Some women are more prone to urinary tract infection, e.g. cystitis. (This may be improved by changing to another type.)
- Not all women feel comfortable touching themselves in this way.

Inserting the cap
- Make sure that it has no holes.
- Place a ribbon of spermicide on both sides of the diaphragm dome and around the rim. (For smaller caps place a tea-spoonful of spermicide into the cap, but *not* on the rim as this might stop the suction.)
- Find a comfortable position and insert the cap by squeezing the rims together and sliding into the vagina.
- Check that it is covering the cervix.

Removing the cap
- Do not remove sooner than six hours after intercourse.
- Using the index or middle finger reach into the vagina and hook the finger under the rim of the cap, pulling downwards.
- Wash with water and mild soap, rinse and dry.

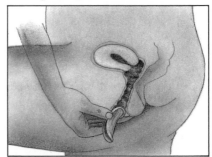

Inserting the cap

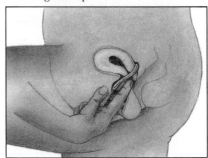

Checking the cervix is covered

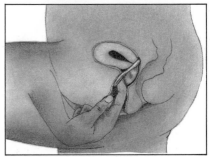

Removing the cap

Spermicides

These are substances that chemically destroy sperm. They are available as jellies, creams, pessaries, aerosol foams and water soluble films. They nearly all contain the same constituent called Nonoxynol 9. They are not effective enough when used alone (seventy-five to ninety-six per cent). Spermicides are not known to have any adverse health risk, although some men and women have allergies to them.

The vaginal contraceptive sponge

This is the newest addition to the range of barrier methods. It is a soft circular sponge, containing spermicide, which comes in one size (5cm across by 2cm deep). It has a small dimple on one side, designed to fit over the cervix, and a removal loop on the other side. It is not generally available on the NHS, but can be bought at any chemist. The sponge works mainly by a continual release of spermicide, and its effectiveness as a 'barrier' is secondary. It is seventy-five to ninety-one per cent effective.

Using the sponge
- Moisten the sponge with water to start the release of spermicide.
- Find a comfortable position and insert the sponge with dimple side uppermost so that it fits over the cervix.
- It remains effective for twenty-four hours, during which time you can make love as many times as wished without adding more spermicide.
- Do not remove sooner than six hours after intercourse.
- The sponge must not be left in place for longer than thirty hours.

Advantages
- The sponges come in one size.
- No specialized fitting is required.
- Use is not related to intercourse.
- There is no mess.
- They are easily available from chemists.

Disadvantages
- There is a high failure rate.
- Some women are allergic to the spermicide.
- They are quite expensive.

The sheath

This is the only reversible method that can be used by men. The main problem is one of image, with sheaths often being regarded as 'passion killers', reducing spontaneity and sensation. Yet they are simple, effective (eighty-five to ninety-eight per cent), ultrafine, multicoloured, multiprotective and can be fun to use. However, the sheath is only as good as the user and therefore needs correct and careful handling.

The sheath is basically a simple latex rubber tube sealed at one end which fits over an erect penis and catches the semen after ejaculation. The majority of sheaths are lubricated with an inert lubricant; some are spermicidally lubricated. There is also a special allergy sheath.

Using the sheath
- Once removed from the packet it should be rolled onto an erect penis *before* any vaginal contact takes place (sperm is present in the pre-ejaculatory fluid).
- The closed end (teat) should be pinched to expel any air. This allows space to receive the sperm.
- After ejaculation, the sheath should be firmly held onto, to prevent it slipping off during withdrawal and spilling semen into the vagina.

Advantages
- Sheaths are very effective.
- They are easy to use and obtain.
- The man can take responsibility for birth control.
- They offer protection against some STDs and cancer of the cervix.

Disadvantages
- Great care is needed in use.
- Using a sheath can interrupt lovemaking, although it can be enjoyed as part of loveplay.

Natural family planning or fertility awareness

Natural family planning (NFP) or fertility awareness (these used to be known as the 'safe period') methods are refined and improved ways of detecting ovulation – the main fertile time in a woman's cycle. Nature has provided women with a highly effective 'built-in' system of fertility regulation, but surprisingly few women realize that it is possible to identify this critical time. Essentially, fertility awareness means becoming aware of the feelings, emotions and bodily changes that characterize the fertile and infertile phases in the menstrual cycle.

The NFP methods
There are essentially four methods and these are:

The calendar method: This involves working out the 'safe' times for sex each month in advance, based on the calculations of six to twelve menstrual cycles. As this method makes no allowance for cycle irregularity, stress, illness etc., it is not a method to use on its own.

The temperature method: This involves noting the temperature changes that occur during the menstrual cycle. Immediately after ovulation, basal body temperature (BBT – the body's temperature at complete rest) drops slightly, then rises by about 0.2°C to 0.4°C. It stays high until the next period. Once this increase is sustained for three or more days, it is considered safe to have unprotected intercourse.

You need a temperature chart and fertility thermometer (this has a narrower range of temperature gradings, so is easier to read than a clinical thermometer). Both are free on prescription. Take your temperature each day, at approximately the same time, as soon as you wake up and before you do anything. Whether you take it orally, vaginally or rectally, always take it the same way. Your temperature can change for reasons other than ovulation – illness, drugs, alcohol etc. – so it is important to record these events on your chart.

The cervical mucus method: This method, also known as the ovulation or Billing's method, relies on recognizing changes that occur in cervical mucus. In the infertile phases of the menstrual cycle before and after a period, the cervical mucus is thick, sticky and cloudy in appearance. Before

Days of cycle

| 1 | 2 | 3 | 4 | 5 | 6 | 7 | 8 | 9 | 10 | 11 | 12 | 13 | 14 | 15 | 16 | 17 | 18 | 19 | 20 | 21 | 22 | 23 | 24 | 25 | 26 | 27 | 28 | 1 | 2 |

Calendar method

safe period	unsafe period (for a woman whose cycle varies from 25-31 days)	safe period

Mucus method

at ovulation mucus resembles the white of an egg

bleeding	dry	mucus increases and gets clearer	mucus decreases and gets cloudy	dry

Temperature method

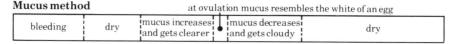

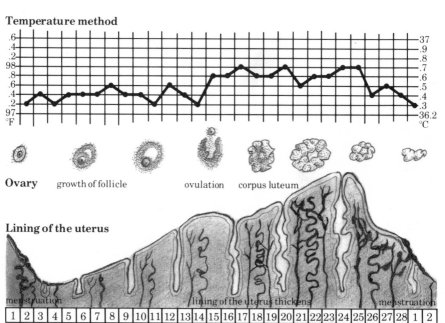

Ovary   growth of follicle   ovulation   corpus luteum

Lining of the uterus

menstruation   lining of the uterus thickens   menstruation

| 1 | 2 | 3 | 4 | 5 | 6 | 7 | 8 | 9 | 10 | 11 | 12 | 13 | 14 | 15 | 16 | 17 | 18 | 19 | 20 | 21 | 22 | 23 | 24 | 25 | 26 | 27 | 28 | 1 | 2 |

and during ovulation (ovulation happens twelve to sixteen days *before* your next period) the mucus changes to become copious, clear and stretchy. The appearance of this kind of mucus marks the fertile period. In the days after ovulation the mucus returns to its thicker state which is naturally hostile to sperm. By noting these daily changes a woman can actually predict when ovulation will occur and avoid intercourse, or she can use a barrier method at this time.

The sympto-thermal method: Also known as the double-check method, this combines two or more of the above fertility indicators. It also includes recognition and interpretation of other changes in the menstrual cycle, such as cervical changes, mid-cycle pain (mittelschmerz), breast sensitivity and mood changes. A woman might experience some, or all, of these changes. For women wanting to avoid pregnancy the sympto-thermal method is the most effective (eighty-five to ninety-three per cent).

Advantages
- Natural family planning has no known physical side-effects.
- Couples share the responsibility for family planning.
- Women gain a greater awareness of their bodies.
- It can be used to achieve or prevent pregnancy.

Disadvantages
- Natural family planning requires the commitment of both partners.
- It requires careful observation and record-keeping.
- Ideally, there should be personal teaching from an NFP teacher. Unfortunately, good NFP teaching is seldom available from family planning clinics or GPs. You can get specialist help from The Natural Family Planning Service and The National Association of NFP Teachers (see Useful Addresses).

Male and female sterilization

Sterilization is the only permanent method of family planning. It is ideal for those who have decided their families are complete, or who are quite sure they never want children. Since many of us do not want to continue using contraception year after year in these circumstances, sterilization is now becoming a popular option.

Because of the permanency of sterilization, counselling should always be offered. This provides an opportunity for any worries, problems or questions to be raised and discussed. The decision to be sterilized should not be made directly after childbirth, abortion or miscarriage or when there is a relationship or personal crisis. If there are *any* doubts about sterilization you should *not* go ahead. Studies have shown very clearly that where a person is happy about the decision to be sterilized there are no regrets or psychological problems.

Operations to reverse sterilization do exist, but the chances of success are not always high and depend on the type of sterilization carried out originally. Methods involving least tubal damage offer the best potential.

Neither partner in a relationship has any rights over the other's body, so consent is not legally required. However, many doctors like to have some assurance from the partner (if there is one) that it is a joint decision.

Male sterilization (vasectomy)

This involves cutting or blocking the vas deferens – the tubes that carry sperm from the testes to the penis. The vas deferens are reached via a tiny cut, either in the middle or on each side of the scrotum. A small amount of tube is removed, or cut and sealed. One in 1,000 failures occur.

A vasectomy is usually performed under local (sometimes general) anaesthetic, as an outpatient procedure – it takes as little as four or five minutes! It is safer and simpler than female sterilization. It has no effects on the male sex organs: testes, prostate, or penis, or the male sex hormone, testosterone. Therefore, it does not affect a man's libido at all. Nothing is taken away except the ability to have children. As the passage of sperm is blocked, the seminal fluid will contain no sperm, but as this makes up such a tiny fraction of seminal fluid, this is not even noticeable. There are no serious long- or short-term effects. The vasectomy can leave a man feeling a little bruised for up to a week, but this can be helped by wearing good support underpants. Swelling and infection can occur, but this is easily treated. Heavy work or lifting should be avoided for a few days.

Sex is possible as soon as it is comfortable, but as sperm take several months to be produced and develop, there will be some left in the tubes after the operation. It takes at least twenty-four to thirty-six ejaculations to clear these sperm. For this reason, two separate sperm tests are taken at twelve to sixteen weeks after a vasectomy. It is important to continue using contraception until two consecutive negative tests occur.

Advantages

- Vasectomy is simple and safe, with no serious long- or short-term effects.
- It is very effective.
- It is not related to intercourse.
- Fear of unplanned pregnancy is removed, so a couple's sex life is often improved.

Female sterilization

There are a number of different ways in which a sterilization can be performed. They all have the same aim, which is to block the fallopian tubes so that egg and sperm cannot meet. The tubes can be reached either

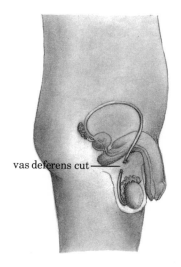
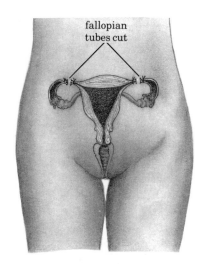

fallopian
tubes cut

vas deferens cut

directly, through a small cut in the abdominal wall – laparotomy or mini-laparotomy – or indirectly, via a tiny incision in which a small narrow telescope-like instrument called a laparoscope is used – laparoscopy. The fallopian tubes can also be reached indirectly through a small cut in the vagina but this is only very occasionally used.

Laparoscopy is the most common method used today, due to its simplicity, speed and minimal inconvenience. Once the fallopian tubes have been reached they can be artificially blocked by tying and removing a small piece of tube (excision and tubal ligation), or by sealing (cauterization and diathermy), or by the use of clips or rings. Female sterilization has a failure rate of one in 300.

The operation can take place either in hospital, or a well-equipped clinic, under local or general anaesthetic. The length of stay depends on the method and anaesthetic used, but it can be as little as one day, without an overnight stay. Sterilization is immediately effective, and sex can be resumed as soon as it feels comfortable.

The operation should not change your sexual feelings, or emotions, or make you feel any less feminine. The womb, cervix, fallopian tubes and ovaries are not touched. The only physical change is that you can no longer have children. It is a safe operation with no long- or short-term serious effects. Some studies have shown that a few women have heavier periods after sterilization, which might lead to hysterectomy. One possible explanation is that, when sealing the tubes with diathermy, the blood supply in the surrounding area is in some way affected. Other research disputes this, indicating that sterilization is too easily blamed for other problems. There is a slightly higher risk of ectopic pregnancy if the method fails.

Advantages

- Fear of unplanned pregnancy is removed, so sterilization can lead to an improved sex life.

- Sterilization is immediately effective.
- It is not related to intercourse.

Morning-after contraception

Even the most organized of us can forget to take the Pill, misplace a cap or simply not use anything at all in the heat of the moment. In fact, since time began a variety of 'after sex' methods have been tried, from vaginal douching to extreme physical exercise, in order to remove sperm from the vagina. Unfortunately, these measures are not effective – sperm are just too well designed and can reach the fallopian tubes within five minutes! However, it is now possible to do something more positive to prevent the possibility of an unplanned pregnancy. Rape victims in particular should be made aware of these possibilities.

The morning-after pill: This is taken within three days of unprotected intercourse and consists of four high-dose oral contraceptive pills (not the usual ones given for contraception). Two tablets are taken immediately, followed by a repeat dose exactly twelve hours later. This method is about ninety-eight per cent effective.

The postcoital IUD: This is fitted within five days of unprotected intercourse and is nearly one hundred per cent effective.

Both methods work by preventing a fertilized egg from implanting. While both are safe, with no serious side-effects, their use must be regarded as a one-off emergency method. As morning-after contraception they should not be considered as a means of regular birth control. Nausea and sickness are the main side-effects for some women taking the morning-after pills. Problems associated with the use of the IUD include pain, cramping and some bleeding and the other recognized side-effects of IUD use.

Birth control in the future

All women wish for a simple risk-free method. Radical new methods such as the male pill, pregnancy vaccines, nasal sprays and 'once-a-month' pills are continually reported by the media as being 'just around the corner'. However, in reality they are still years away. With the transition from a new idea to the final product taking up to twenty years, even things we see on the horizon may never arrive.

Most of us realize that there is no perfect method. However, with careful choice and use, the current range of methods should provide effective and safe protection against pregnancy, and at the same time contribute to a happy and healthy sex life.

6 Childbearing

by Heather Bampfylde

For many women, pregnancy and childbirth are a voyage of self-discovery – when they get to know their bodies and how they function for the first time in their lives. However, many complain that instead of it being a happy and fulfilling experience, they are denied accurate information on which to make the best decisions for themselves and their babies, and also the kind of birth they want. When you are pregnant, it is your *right* to choose the kind of medical care and delivery you wish to receive, and your *duty* to discover the options available and to discuss them not only with your partner but also with the medical professionals – doctors, midwives and consultants. This process starts as soon as your pregnancy is confirmed.

How do you know that you are pregnant?

The first signs are missing a period followed by tell-tale tenderness in the breasts and passing urine more frequently than usual. Wait two weeks from the missed period date before taking a pregnancy test. A negative result may occur if it is undertaken too soon, even though you may actually be pregnant. You can go to your doctor or clinic or buy a home testing kit. If you do the latter, follow the instructions carefully. The advantage of home testing is that you get a quick result without any of the waiting and anguished anticipation of laboratory testing (see Chapter 7). But if the result seems unclear for any reason, you will need to visit your doctor anyway. Take a urine sample with you from the first stream you pass in the morning. This is tested for the presence of the hormone human chorionic gonadotrophin (HCG) which only occurs about twenty days after conception.

Your doctor may examine you and diagnose pregnancy from an internal examination (enlargement and softening of the uterus) and from the appearance of Montgomery's tubercles (enlarged bumps) on your breasts. As soon as your pregnancy is confirmed, arrangements will be made for you to be booked into a hospital for delivery and you may be asked to make decisions about the type of birth you want even before you have grown accustomed to the idea of being pregnant. Now is the time to start asking questions and to obtain all the relevant information possible so that you can make an informed choice.

Clinic visits

In Britain, ninety-eight per cent of women deliver their babies in hospital, and therefore the chances are that an appointment will be made for you to attend a clinic to book you in for the birth. Even though you may be sent a letter or given an appointment card with a specified time, you must prepare yourself for a long wait. Experiences vary but it is not uncommon to spend the best part of a morning or even a day waiting to see doctors, midwives and consultants. It pays to ask for the earliest appointment possible and then to arrive in good time to ensure that you are one of the first women seen before a backlog builds up as the day progresses.

You will be interviewed by a nurse, probably a midwife, and will be asked to give details of your medical history. She will try to determine accurately your EDD (estimated delivery date) from the information you give her about the timing of your last period. Don't be afraid to talk frankly about your medical history and sexual health – it is important that the hospital should know all the relevant facts. For example, if you have ever had an abortion, or if you or your partner has had herpes. These may affect the management of your pregnancy and the birth of your baby.

The interview usually takes the form of an informal chat, and it is your opportunity to ask about hospital policy on birth, pain relief, natural labour, whether they have birthing stools etc. Do not be afraid to admit your ignorance and insist on getting a clear picture of what lies ahead of you. One of the problems involved in establishing a good 'working' re-lationship with your doctor or midwife in a large hospital's clinic is that you may see different people on each visit you make. Even though you will be allocated to a particular consultant obstetrician you may not even meet him or her very often.

In addition, you will be asked to take a urine sample with you for testing for protein (a symptom of pre-eclampsia or kidney problems), glucose (may occur in pre-diabetic conditions) and bacteria (a sign of possible urinary infections). Your blood-pressure will be taken, and also a blood test to establish your blood group and immunity to German measles. You will be weighed and measured and may be asked for your shoe size. This may sound puzzling but the length of a woman's foot helps indicate the size of her pelvis and her chances of a normal delivery.

Most women are asked to visit the clinic for a checkup once every four weeks until twenty-eight weeks into their pregnancy, and then once every two weeks until the thirty-sixth week. Thereafter it is once every week until birth.

Your co-operation card
You will be given a co-operation card on which your pregnancy details are

charted. Your medical history and the results of tests and all screening will be noted down on this card. You must take it with you each time you visit the hospital, clinic or your GP. In fact, as your pregnancy progresses it is a good idea to keep it with you at all times – just in case you go into early labour and information is needed speedily.

At first sight it will look like a mumble-jumble of abbreviations and symbols, and it will help you to understand your own state of health and your baby's development if you know what they mean:

AF amniotic fluid

AFP alphafetaprotein (to check for spina bifida and Down's syndrome)

ARM artificial rupture of membranes

BP blood-pressure

BPD biparietal diameter (baby's head size measured by ultrasound)

Ceph cephalic (head first)

CS Caesarean section

DLMP date of last menstrual period

EDC/EDD expected date of confinement/delivery

Eng engaged (when baby's head drops into pelvis)

FBS fetal blood sample

Fe iron

FH/FHH fetal heart (heard)

FHR fetal heart rate

FMF fetal movement felt

Fundus the top of the uterus

Hb haemoglobin

H/T hypertension (high blood-pressure)

MSU mid-stream specimen of urine

MW midwife

NAD nothing abnormal detected

NE not engaged

N/P not palpable (cannot be felt)

NTD neural tube defect (spina bifida)

OA occipito anterior (baby head first facing mother's back – best birth position)

Oedema swelling

OP occipito posterior (baby head first facing forwards – may create backache in labour)

OT occipito transverse (baby head first but sideways)

PE/VE vaginal examination

PET pre-eclamptic toxaemia (not serious – symptoms are rise in blood-pressure but no danger if treated quickly)

Rh Pos Rhesus positive blood group

SVD spontaneous vaginal delivery

U/S ultrasound scan

Tests during pregnancy

The two principal tests for assessing the health of your baby are ultrasound scanning (often carried out routinely in the fourth month of pregnancy) and amniocentesis (offered only to women over thirty-five or so depending on the individual case history and hospital policy).

Scanning

An appointment will almost certainly be made for you to have a 'scan' and you will be asked to drink a pint of water before you attend – the scan picture is clearer with a full bladder. You may be able to take your partner along, too, for your first sight of your baby. There is nothing alarming about being scanned. Your abdomen will be oiled and a metal arm is moved across it by the operator. High-frequency sound waves are bounced off the baby and appear as dots on a television-type screen. You may find it difficult to discern the shape of the baby at first but the operator will explain the picture to you and you will start to recognize legs, arms and the head. Scanning is carried out for many reasons:

- To determine the estimated date of delivery (by taking measurements of the baby's head)
- To assess the baby's health (fetal movement and beating heart)
- To assess the placenta (its general condition and position especially in late pregnancy)
- To determine the baby's position in late pregnancy
- To chart the baby's growth and development

Scanning seems to be a safe procedure but doctors admit that they do not yet know whether there are any long-term ill effects to the baby – for example, to its hearing in later life. At the moment, the benefits certainly seem to outweigh any potential long-term risks.

Amniocentesis

Older mothers may opt for this test which can be used to detect about forty abnormalities in the fetus, including neural tube defects (spina bifida) and Down's syndrome. There is a significantly increased risk of having a baby born with a congenital malformation or handicap in women over the age of thirty-five, and, if an abnormality is found, a woman may be offered a termination.

However, although it seems sensible to have this test if you are in your late thirties or early forties, you should be aware that there is a one to two per cent chance of the baby aborting as a direct result of the test. Your doctor should inform you of this danger and you must discuss it with your partner thoroughly before you go ahead.

After a painkilling injection of local anaesthetic, a needle is inserted through your abdominal wall and a sample of amniotic fluid containing cells from the baby's body is withdrawn for testing.

Talking to your doctor and midwife

If your doctor regards pregnancy as a time of enforced rest and idleness, you may be infuriated by his/her attitude, especially if you are very active and intend to work throughout your pregnancy. Do not be discouraged unless there is a good medical reason for you to stop and put your feet up. Likewise, if your doctor never addresses you face to face but tends to talk to your stomach or to the midwife or nurse present, or refuses to treat you as an intelligent thinking person who wants to understand what is happening to you and your baby and make choices for yourself.

Most women are generally happy with their relationships with their doctor and midwife, but some complain of being treated as helpless and dependent. Some doctors may expect them to be passive and unquestioning and speak condescendingly to them. Insist that your doctor treats you as an intelligent person, supplies you with the information you need to know and gives you reasons and valid arguments for any instructions or courses of action he suggests.

Often a woman finds that she has a better relationship with her community midwife or, if she is lucky enough to meet her in advance, the midwife who will deliver her baby. She may be more sympathetic than a doctor and have more time to talk and to answer questions.

Women who encounter problems may opt for a home delivery with a midwife in attendance rather than a hospital birth. Remember that it is your right to choose. If you are very unhappy about the advice you are receiving from your doctor, it is possible to change to another doctor. Do not be afraid of making a fuss if you believe that you are making the right decision. See Chapter 16 for advice on how to go about this.

In the same way, if you are worried about the hospital into which you have been booked for delivery, find out about other facilities in your area and talk it over with your doctor or midwife. It should be possible to change to another hospital or birth environment.

Women over thirty may encounter special problems in their dealings with the medical profession. For a start, it is not very flattering to be described as an 'elderly primagravida' when you still feel young and attractive. You may be treated as though you are rather old and fragile when you feel in the best of health, energetic and robust. More and more women are now opting to have their children later, in their thirties, after they have established themselves in a career and found some financial stability. They are becoming more the norm than the unusual, and medical personnel must adapt to this new way of thinking.

Choosing the birth you want

During the months leading up to your delivery you will have time to find out about different methods of birth, those that are practised in your area and the sort of birth you want. Ask your doctor and midwife for information, read as many books as you can and talk to others who have had children about their experiences so that you can make an informed choice.

Home or hospital
The first decision to make is whether to have your baby at home or in

hospital. Both have pros and cons but the chances are that you will be encouraged to have your baby in hospital, especially if it is your first child, if you are over thirty or if you have had complications in previous births or pregnancies. Some women prefer to deliver at home in familiar surroundings with their partner, a friend or member of the family helping. It is comforting and less intimidating than the sterile atmosphere of a large hospital and a high-tech birth. You are entitled in law to do this and the local authority has a duty to provide a midwife. However, if any complications were to arise, it would be necessary to rush you to the nearest hospital for specialized treatment with subsequent risks to your health

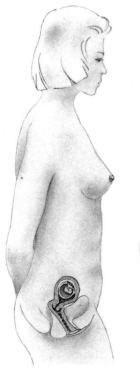 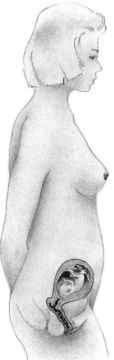 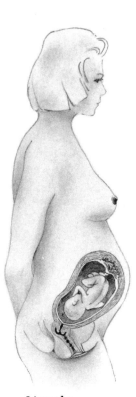

8 weeks
The embryo at eight weeks begins to look more like a human being. Limbs have developed and all the main organs have formed. The fetal heart beat is strong. The mother may be experiencing sickness and her breasts will be enlarging.

16 weeks
At this stage it is possible to tell the sex of the baby. Its head is still large for its body and its transparent skin allows the blood vessels to show through red. The mother's abdomen has begun to bulge and feelings of sickness have probably stopped.

24 weeks
The fetus has now grown so that the mother is obviously pregnant. Fat deposits begin to fill out the baby's wrinkled skin, which is also losing some of its redness. The mother may be experiencing backache as the muscles stretch.

and that of the baby. Whereas in hospital it is reassuring to know that there are always skilled, highly trained medical staff on hand to cope in an emergency and offer pain relief if needed. A paediatrician can examine the baby immediately after birth and you can learn how to feed and look after a first baby and have a rest before you return home.

Natural birth or drugs

Most women want a natural labour free from drugs and medical intervention. They want to be conscious and in control and fully able to experience this fulfilling and momentous time in their lives. However, the

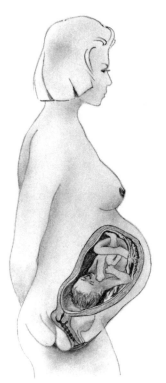 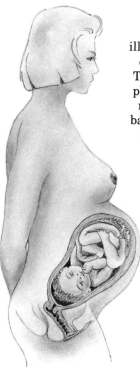

Pregnancy
The five stages of pregnancy shown here illustrate the development of the fetus in the womb. They also chart the major physical changes that the mother undergoes as the baby grows and gradually positions itself for birth.

32 weeks
The lungs have developed and the baby would have a good chance of survival if born at this stage. In response to loud noises, the baby will probably kick and move its arms. Increased progesterone levels, which help prevent premature labour, may mean that the mother lacks energy.

40 weeks
The pregnancy has reached full term. The baby's skin has smoothed out and the head, which is now more proportionate to the body, may be covered with hair. The baby's eyes are open and the facial features well formed. Also well formed are the baby's genitals. The mother awaits with anticipation the birth of her child.

best intentions are sometimes thwarted by a very painful delivery, especially if you have a backache labour, and you should explore the types of pain relief available beforehand and be prepared to accept drugs if you feel you need them. You should not feel guilty if you opt for pain relief in labour – there is nothing wrong with preferring a pain-free delivery. Pain can be tiring and debilitating and may mar the moment of birth and holding your baby for the first time. Keep an open mind and react according to your instincts when the moment arrives to make a choice. All drugs, except an epidural, enter the blood-stream and pass through the placenta into the baby and may affect it to varying degrees, although there are no long-term risks to health.

Types of pain relief

The most commonly used methods are:

Gas and air: A woman administers this herself by holding a mask over her face and inhaling deeply and slowly. However, it can make you feel heady and it does not bring relief to everyone. In addition, the inhalations have to be timed carefully and practised if it is to be effective.

Pethidine: This is administered by injection to reduce pain but it may induce drowsiness.

Anaesthetics: Although these certainly do dull the pain, they may cause you to lose consciousness altogether, or at least not to be fully aware of the birth of your baby.

Epidural: This is generally considered by many women to be the most effective form of pain relief as it enables you to remain fully conscious. Only a qualified anaesthetist can administer the injection of local anaesthetic into the space in the spinal column that surrounds the nerves to the legs and abdomen, thus deadening the lower half of the body. Its disadvantages are that you will be put on a drip to control your blood-pressure, you may feel nauseous and it will be difficult to push the baby out. Most epidurals necessitate a forceps delivery and episiotomy.

Caesarean versus vaginal birth

If your doctor recommends a Caesarean section, ask him/her to explain the reasons for this. They may include a breech delivery; a low birthweight baby; the baby's head is too large to pass through your pelvis if you are very small-boned; or if you have diabetes, chronic hypertension, severe back problems or genital herpes. If you feel that it is not strictly necessary as in the case of a breech birth, do not be afraid to argue the point with your doctor and ask for a second opinion if necessary.

A Caesarean may be performed under general anaesthetic or epidural anaesthesia. The latter is generally preferable and safer, too, as you are

Right: A pregnant woman gives herself gas and air to relieve the pain of labour. Not all women find that this method works, but with careful timing and practice it can be effective. *Below:* The moment of birth with the baby's head turning sideways as it enters the outside world. The mother has opted for the squatting position to deliver her baby.

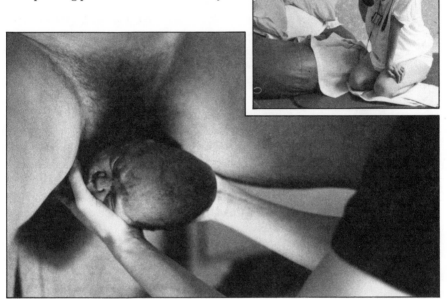

fully conscious throughout and the baby is unaffected by the drugs. The operation takes about four minutes – an incision is made horizontally below the bikini line and the baby is delivered abdominally. There is no need to feel a failure or as though you have missed out on a marvellous birth experience just because you do not undergo a full labour and vaginal delivery. Many women are happy to have a Caesarean section and miss out on the pain of labour and the discomfort of episiotomy stitches afterwards.

Of course, there is a painful wound but you will be able to get up within twenty-four hours of the delivery. It will take about six weeks to heal completely and the scar will gradually fade.

Natural birth

This means a natural delivery without anaesthesia or pain relief in the form of drugs. Now that most women express a preference for this form of birth, many hospitals are becoming more enlightened and may promote it if you are fit, healthy and undergo a trouble-free labour. When you arrive, you may be offered a warm bath, a rocking chair or birthing stool. Walking around in the early stages of labour and even standing up or kneeling on all-fours to give birth may be actively encouraged if this is what you want to do. The atmosphere may be deliberately homely, more comfortable and less sterile than you expected; the lights may be dimmed for a gentler entry for your baby into the outside world and you will be discouraged from taking drugs unless you are absolutely set on having them.

If you want a natural birth, it is essential that you attend antenatal classes and learn how to relax and practise the breathing exercises. You can control your labour better if you can consciously regulate your breathing during contractions and relax to reduce tension and ease pain. Do not be fooled into thinking that breathing can prevent pain – it can't. However, it can make it easier to bear so that you are more aware of the birth of your baby and in charge of your body and what is going on.

Antenatal classes will help you prepare mentally and emotionally as well as physically for labour and the new baby. Many women find them reassuring and supportive and a good opportunity to find out more about the options open to them and get their questions answered. A woman should start attending these classes from the fourth or fifth month onwards. They are free and are sometimes held in the evenings as well as during the day. If you cannot get along for daytime sessions, you could enrol in a class run locally by the National Childbirth Trust. NCT classes not only help you cope with an active birth but they also teach you about breast-feeding your baby and you can meet your local breast-feeding counsellor who can advise you afterwards. See the addresses at the back of the book and write to the NCT for the name of your local representative.

To induce or not...

There used to be a lot of horror stories about induction being a matter of policy in many hospitals but thankfully this is no longer the case and few hospitals induce women routinely to reduce staffing problems or better control a normal labour.

An induced labour is started artificially, either by administering a pessary or an intravenous drip of oxytocin. Labour may be more painful and much faster than it would be otherwise, and if your doctor suggests induction he should always be able to back it up with sound medical arguments. There is no reason why you should allow yourself to be induced unless you have a medical condition that necessitates it and there is a risk to yourself or your baby by prolonging the pregnancy further.

An 'overdue' baby is often cited as the reason for induction, but it is possible to get dates wrong when estimating the EDD, especially if you have had irregular periods. A normal pregnancy can last anything between thirty-eight and forty-two weeks which gives plenty of scope for mistakes in calculations. If it can be shown that the baby really is overdue and the placenta is no longer functioning efficiently in supplying the baby with the required oxygen and nutrients, then it will be necessary to induce labour. Do not be steamrollered into a decision on the spot if you have any doubts – discuss it with other medical personnel, your partner and family first.

Sex during and after pregnancy

There is no need to say goodbye to your sex life just because you are pregnant. Right up to the last days before the baby is born you can continue to enjoy intercourse. During the first trimester (three months), nausea and morning sickness may interfere with your lovemaking and lead to a reduced or changed routine. However, many women claim that it is more satisfying than ever before, in the second trimester especially.

Lovemaking cannot harm your baby, safe within the amniotic fluid-filled sac in the uterus. However, you may have to experiment with different positions and avoid the traditional 'missionary' position with the man on top. Positions where the woman is on top or on all-fours, or both man and woman lie on their sides facing the same way are all suitable and cannot harm your baby. Sex cannot trigger off labour unless the baby is due, in which case the prostaglandins present in the man's semen may cause it to start.

It may take time to re-establish your sex life after the birth of your baby, especially if you have had stitches. Wait two or three weeks at least for the episiotomy wound to heal and be very gentle while you build up your confidence. You may need to use a lubricating jelly as scar tissue may cause discomfort or even pain. Try to relax, cuddle up and go very slowly. The seated position with the woman on top of the man will probably be most comfortable initially, especially if your breasts are swollen and full of milk. If you feel sore afterwards, have a hot bath to ease out tension. If you have persistent pain during intercourse, you should ask your doctor to examine you. It may be a relatively simple matter to eliminate the problem.

Of course, if you were lucky enough not to tear or have stitches, you should be able to resume sex quite quickly – as soon as you feel ready. Many women find that their sex lives are more enjoyable and that they have a closer, more loving relationship with their partner after having a baby.

7 *Unplanned Pregnancy*

by Antonia Rowlandson

The first signs of pregnancy such as a missed period, nausea or tender breasts can be a great shock to a woman who has been trying to avoid it. However, if she has taken a risk or contraceptive measures have failed, it is important that she finds out as soon as possible whether she really is pregnant, rather than waiting and hoping that the problem will go away on its own. If she decides that termination of the pregnancy (abortion) is the best course of action, then she should try to arrange this as soon as possible. Terminations carried out in the early weeks of pregnancy are better for the woman from an emotional and medical point of view. The operation itself is a very common one – 170,000 pregnancies are terminated in England and Wales each year, 35,000 of these operations are performed on women from abroad.

Pregnancy testing

A specimen of urine taken first thing in the morning can be tested in a laboratory or clinic for human chorionic gonadotrophin (HCG), a hormone that is secreted as soon as a fertilized egg becomes implanted in the lining of the womb. About an eggcupful of urine should be collected in a clean container free from soap or detergent and labelled with name, date the specimen was collected and date of the first day of the last menstrual period. A positive result means that there is a strong possibility that the woman is pregnant – false positives are rare. But if the result is negative, she may be advised to have another test a week later to make sure that she is not pregnant – hormone levels in the urine may not be high enough to

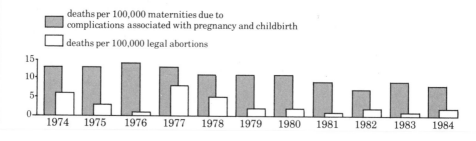

deaths per 100,000 maternities due to complications associated with pregnancy and childbirth

deaths per 100,000 legal abortions

diagnose pregnancy if the test is done too early. After sixteen weeks a urine pregnancy test may not work, and pregnancy should then be diagnosed by physical examination. A blood test can also confirm pregnancy.

There are also several kits available from chemists that can be used at home to test urine for signs of pregnancy. Recently developed home testing kits, such as Clearblue and Predictor Colour, incorporate a new monoclonal antibody and enzyme-based technique that makes testing much simpler and quicker, and potentially more accurate. The test is very sensitive and so can be used from the first day of the first missed period. These new tests are claimed to be ninety-eight to ninety-nine per cent accurate but the reliability of the test will always depend to some extent on the care with which it is carried out. Some family planning and pregnancy advisory organizations can carry out these very early tests, as well as the standard urine test, but they tend to be more expensive. In general they will only be offered on the NHS under special circumstances when a woman needs the test result urgently.

In some areas lack of funds has meant that there have been cutbacks in the availability of NHS pregnancy tests. GPs have to pay for test kits themselves as the NHS does not provide them. Therefore, most GPs send women to a hospital laboratory rather than do the tests themselves. Family planning clinics will only carry out free testing on women who are registered with them. If a woman does not want to wait, the alternative is to pay for a test (for example, this costs about £3.00 at the British Pregnancy Advisory Service) or to buy a home test kit – these cost about £6.00. Also, some pharmicists in high street chemists will carry out pregnancy testing, and may be able to advise on the use of home test kits.

Where to go for help

Once a woman knows for sure that she is pregnant, she should discuss the situation with her partner. It is worth remembering that even though a pregnancy may not have been planned, it is not necessarily unwanted. A woman may not have thought very deeply about having children, but once she finds that she is pregnant, it may turn out that she and her partner agree that they want a child, and that this would be a good time to have one. At the other extreme, a woman who is sure that she does not want to be pregnant but who has taken a risk and thinks that there is a strong possibility that she might be pregnant can opt for 'morning-after' or 'post-coital' contraception (see Chapter 5).

For others, however, the situation may not be so simple, and discussion with partner, other members of the family or friends may not be enough to help a woman decide what to do next. Alternatively, she may not *want* to talk to those closest to her, at least until she is sure about what she is going to do. In this case she will need some outside help, and her GP could be the person she turns to first. If for some reason a woman cannot talk to the GP about this problem – perhaps he or she is an old friend of the family, or strongly opposed to abortion – then she can either change to another GP or get advice and help from the local family planning clinic, Brook Advisory Centre (mainly for those under twenty-five), Marie Stopes Clinics, Margaret Pyke Clinic or an abortion agency such as the British Pregnancy Advisory Service (BPAS) or the Pregnancy Advisory Service (PAS). These organizations will also carry out pregnancy testing.

Looking at the options

Once a woman knows for sure that she is pregnant, she has three basic options: she can have the baby and keep it; she can continue with the pregnancy but arrange for someone else to look after the baby, perhaps have it adopted; or she can have the pregnancy terminated.

Deciding to keep the baby
This is a particularly big decision for a woman who is unlikely to have the support of her partner. Under these circumstances, she will have to be very honest with herself about the practical aspects of caring for a child on her own. Can she afford it, where will she live, what about her work, the qualifications she hopes to get, or her schooling, who will look after the child while she works, can she rely on her parents or friends to help her out? – these are some of the questions she will have to ask herself. Organizations such as Citizens Advice Bureaux, Local DHSS offices and the Child Poverty Action Group should be able to help with housing rights and entitlements to state benefits.

Also, looking after a baby is not like a job or a relationship which can be dropped at any time. Children are an occupation lasting all day, every day that goes on for years – some would say for ever. The stress and responsibility of looking after a child single-handedly is probably greater for the very young mother. If the father of the baby will not be staying with her or marrying her, but she likes the idea of another relationship or marriage, how difficult will she find it to get out and start a new friendship? Some young men may be reluctant to embark on a serious relationship with a girl who already has a child.

Of course many young, unmarried mothers cope very well with their children and enjoy them enormously; they go on to marry and have more children. However, it is still very important for a woman to look realistically at the situation and to appreciate how having a baby may affect those around her, such as her parents, before deciding to continue with the pregnancy. There are several organizations that specialize in giving help and advice to young, single mothers or single pregnant women. These include One Parent Families, Gingerbread, and Life, an organization strongly opposed to abortion that provides counselling and support for single pregnant women and mothers.

Women who already have children may be in a rather different situation if they become pregnant by mistake. Sometimes they find themselves under more pressure to continue with the pregnancy than women without children. At other times, although they may already have a home and a stable environment in which to bring up a child and may be more aware of what is involved, they and/or their partner may feel that they have all the children they want or can afford, or perhaps they were hoping for a gap before having more children. Some women may want to reach a decision without telling their partner, but discussing the situation together should help a couple reach the best decision for their particular circumstances. And of course they can also seek the help of their doctor, marriage guidance counsellor, or an abortion charity.

Having the baby cared for by someone else
Some women, particularly young women, who feel they cannot cope with a child but do not want to have an abortion, hand over the care of their baby to someone else. It may be possible for their mother or aunt or a friend to look after the baby, particularly if she still has young children at home and can therefore care for the child as another one of her own. However, the child's biological mother may find this situation difficult, especially if she sees the child regularly, since she will have to give the new carers total responsibility for the child and will have to restrain herself from interfering if the child is not brought up exactly as she would like. Therefore, some women may find legal adoption a more satisfactory way of handing the child's upbringing over to someone who may be able to offer it a better life.

Adoption can be arranged by local social services departments or by an adoption agency. A woman is never rushed into handing over her child if she is considering adoption. Many who have thought about adoption when they are pregnant change their minds once the baby is born. A child can be put into foster care while the mother makes up her mind and she will not be able to give final consent to adoption until at least six weeks after the birth. Adoption can be very traumatic for the mother and she is unlikely ever to forget what she has done, but it is reassuring to know that adoptive parents are rigorously vetted (these days there are many more parents wanting to adopt than there are available babies) and the mother will have some say in what sort of family she would prefer her child to be brought up in. If she really feels she cannot cope with a child, adoption may be the surest way of giving her child the best possible chance of a good home.

The child's biological mother will have no contact with the child once she has given legal consent to the adoption. However, once the child is sixteen he or she can see the original birth certificate, find their mother's name and address and trace her if they want to.

Terminating the pregnancy

Some women will have strong religious or ethical reasons for not considering this option and will therefore continue with the pregnancy and make the best arrangements they can for the child once it is born. However, for those who do see abortion as one of the options, this is still not an easy decision to take; similarly, women opposed to abortion in principle often re-consider the options open to them when faced with an unplanned pregnancy.

In the days and weeks during which a woman tries to decide what to do about an unwanted pregnancy, she will experience many different emotions and feelings. It is likely that her feelings towards the pregnancy and the father of the child will swing from one extreme to another. Perhaps to begin with, when she is still in a state of shock after finding that she is pregnant, she will feel sure that an abortion is the only solution, the best way of wiping the slate clean and carrying on as if nothing had happened. But as the days go by she may begin to imagine what it would be like to have a child. Looking around her at the small children of friends, a child of her own may seem a rather appealing idea, particularly if she has no other definite plans for the future. However, the reality of being tied down to a small child may then sink in, and she may begin to blame herself, and possibly her partner, for being in this apparently impossible situation.

Feelings of anger and guilt may also be strong at this time. A woman may feel that it was all her partner's fault, or she may be angry with herself for making a mistake and getting pregnant, or she may feel ashamed and guilty about the encounter that resulted pregnancy. This is a lot for a woman to go through, particularly when she may not be feeling her best as she experiences the symptoms of the first stages of pregnancy such as nausea. But it is important that she has the chance to examine her feelings, even though this may mean talking and thinking about subjects such as relationships and plans for the future that she has hardly considered before.

Professional counsellors at abortion charities or other organizations will do as much as possible to help a woman explore her feelings but at the end of the day it is up to her, and possibly her partner as well, to come to a decision about termination. Although it is important for a woman to confirm pregnancy as soon as she suspects it, and to start thinking about what she is going to do, it is not wise to make a hasty decision. Whatever her future feelings are about the decision she makes at this time, at least she will be reassured by the fact that she made every attempt to make the best decision with all the knowledge and help that was available to her.

Where to have an abortion

Abortion is legal in England and Wales if two doctors certify that continuing with the pregnancy would be a risk to the physical or mental wellbeing of the woman or her family, or if there is a risk of the baby being abnormal. This means that a woman who is distressed by an unwanted pregnancy, perhaps because of the social circumstances, can have an abortion. However, some doctors are opposed to abortion and they are allowed to refuse to help a woman have one unless the pregnancy puts her life at risk. If a woman's GP is reluctant to help her have an NHS abortion, or if she prefers not to see him, then she can go to another doctor, family planning clinic or similar organization where she will be helped with referral for an abortion through the NHS. Alternatively, she can go to one of the abortion agencies, or Marie Stopes, from where she will be referred to a private nursing home if she wants an abortion. This will cost approximately £150–200, depending partly on how far advanced the pregnancy is at the time of abortion. The other option is to go to a private abortion clinic which is not attached to a charity. All nursing homes that provide abortions and all agencies that counsel and take fees are registered with the DHSS.

Daycare abortions

Some clinics, private and NHS, carry out abortions on a daycare basis. This means that the woman, who usually though not always has a local anaesthetic, can go home on the same day as the operation. Not all areas offer this service but it is worth enquiring about it. There are certain conditions that have to be observed before a woman can be considered for this kind of abortion: she must be less than three months pregnant, she cannot be discharged less than three hours after the operation, the journey home must not be more than two hours or fifty miles and she must have someone to take her home and to stay with her overnight. In addition, she must have the agreement of a doctor in her area to provide emergency cover overnight.

Private or NHS?

The availability of NHS abortions varies greatly from area to area, depending to some extent on how keen local doctors and consultants are on abortion and the limit they set on the number of operations performed each week. Although on average just over half the abortions done in this country are carried out privately, in some areas as few as ten per cent of abortions are done on the NHS, whereas in other areas most abortions will be on the NHS.

If a woman finds that an NHS abortion is difficult to obtain in her area, it might be a good idea to get help from an abortion charity to prevent any more unnecessary delay. Once a woman has made up her mind not to continue with the pregnancy, the sooner she has the operation the better. If she has trouble paying for it, the charity may be able to offer a payment plan, or in special cases may be able to help arrange an NHS abortion.

The main advantages of a private abortion are that there will be less delay in arranging it and the doctors working in private clinics carry out more of these operations and are therefore more experienced than a gynae-cologist in a general hospital who does many other different operations as well. These factors may explain why complications such as infections and excessive bleeding after the operation are less common in the private sector, where counselling and aftercare are also better. Similarly, there tends to be a more supportive atmosphere in private clinics where the staff are trained to care for abortion cases and where, of course, the wards are not mixed gynaecological wards as in NHS hospitals.

What happens in an abortion operation

Most abortions are carried out within the first twelve weeks of pregnancy and, as abortion is much easier to perform in the early stages of pregnancy, it is important to try to arrange an appointment for the operation within this period. Going into hospital for an abortion can be frightening and upsetting, but there is very little to worry about as it is an extremely safe procedure and is also very common.

Terminations before 12 weeks
Vacuum aspiration is the most common method. This involves dilation of the cervix and evacuation of the contents of the uterus by very gentle suction, and may be done quite easily under local anaesthetic although

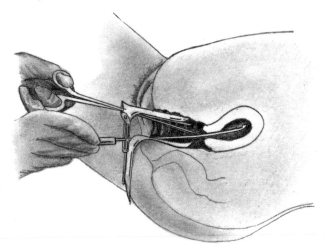

Vacuum aspiration is the most common method of abortion and is carried out before 12 weeks. The cervix is dilated and the contents of the uterus are evacuated by very gentle suction.

general anaesthetic may be used. The main advantage of local over general anaesthetic is that it has far fewer side-effects. The operation is all over in ten or fifteen minutes and the only effects, apart from the side-effects of the general anaesthetic if this is given, may be painful cramps like period pains and some bleeding. But it is important to see a doctor if heavy bleeding or pain continue. Some of the contents of the womb may not have been expelled and a D and C and possibly antibiotics may be necessary.

Terminations after 12 weeks

A lot of patients in this category have, in fact, been kept waiting by the NHS; others may be young girls or menopausal women who did not realize they were pregnant. Abortion may be more difficult at this stage, but it may, nevertheless, be necessary for a woman who did not realize she was pregnant for several months or who needs a termination because her baby appears to be abnormal. However, it is still a very safe procedure. Statistically, pregnancy is much more dangerous than abortion, but in both the risk of serious complications or death is very, very small.

Amniocentesis: The contents of the uterus are expelled by inducing uterine contractions. This can be done by introducing saline solution or prostaglandins into the uterus. The contents of the uterus may not be expelled for several hours but the woman can be given sedatives to make the experience more bearable. Cramps and bleeding may last for several days after the operation but everything should be back to normal by the time of the final checkup six weeks later.

Dilatation and evacuation (D and E): Increasingly, this technique may be used to terminate a pregnancy after twelve weeks. Under general anaesthetic the cervix is dilated and the contents of the womb are broken up and then removed. Some clinics, such as Marie Stopes, prefer this method to prostaglandin induction since it avoids the need for the woman to go through the contractions associated with labour.

Legally, abortions can be carried out up to the twenty-eighth week of pregnancy, but in practice the operation is rarely performed after the twenty-fourth week.

After the termination

Abortion is not dangerous and if it is properly carried out and no complications develop there is no reason why it should affect fertility or future pregnancies. It is, nevertheless, a surgical experience that carries some risk, and one to be avoided if at all possible – no one should ever think of abortion as a method of contraception. So although sex may be the last thing a woman feels like after an abortion – to begin with there may be painful associations between any sexual activity and her experiences of the last few weeks – she should look seriously at the adequacy of her usual method of contraception, if any. Most hospitals or clinics where abortions are carried out will give a woman contraceptive advice after an abortion and she should ask for information if it is not given. It is important to be aware that a woman *can* get pregnant before her first period after an abortion.

Many women feel extremely relieved after the abortion has been carried out. They have probably spent weeks worrying about whether they were making the right decision and then dreading the operation itself. But at last all that is over and they can start getting back to normal again. Others may be more affected by depression, guilt or anger. However, everyone reacts differently and feelings after the event will depend to a large degree on the circumstances of the abortion. A teenage girl, possibly still at school, will be relieved to get back to friends, school and work and may not have much time for guilt or sadness. On the other hand, a woman who already has children may feel bad about not having the child since she will be more aware of the significance of what has happened. She may guiltily think that perhaps she could have coped with it after all.

The continued support of partners, family, doctors and other professionals is vital at this time for those who have difficulty in putting everything behind them. If a woman keeps her feelings to herself she may find that it takes her much longer to get over the experience and she may have difficulties with future relationships and pregnancies. However, if a woman has had a chance to talk over her problems and to make up her own mind about termination, and receives good aftercare – she can always go back to her doctor or abortion agency if she needs more help in understanding her feelings and accepting the situation – then she should make a full recovery. And the whole experience may have the positive effect of bringing a family or couple closer together. Unwanted pregnancy and abortion may also encourage a woman to take greater responsibility for herself which should help her to feel that she has more control over her life in the future.

8 Miscarriage

by Antonia Rowlandson

Miscarriage means that a pregnancy spontaneously comes to an end before the fetus is able to survive outside the womb. The medical term for miscarriage is 'spontaneous abortion', in contrast to 'induced abortion' which is the term used for deliberate termination of pregnancy, commonly just called 'abortion'. The incidence of miscarriage is not known accurately because many pregnancies end in the first few weeks after conception before a woman even knows that she is pregnant – the only sign may be a rather late, heavy period. However, it has been estimated that between fifteen and twenty per cent of diagnosed pregnancies end in miscarriage and that probably a much higher percentage of early stage, unconfirmed pregnancies also end spontaneously.

Many women have one miscarriage and then later go on to have a trouble-free pregnancy producing a healthy baby. But one in five women who have a miscarriage will miscarry again in the future and it is these women who need to be investigated to try to establish a reason for the problem.

There are many different causes of repeated miscarriage. Some appear to be nature's way of getting rid of an imperfect fetus, others may occur in a woman who is not physically fit to produce a healthy infant – she may have diabetes or heart or kidney disease. And it has even been suggested that in a few cases repeated miscarriage may occur in a woman who deep down doesn't want a baby. Although miscarriage is a common event, there is still much that is not understood and doctors cannot always find a cause. Whatever the reason, there is no doubt that miscarriage is a frightening and disturbing experience and that sympathy and support from relatives and doctors is vital if a woman is to make a full recovery and embark on a successful pregnancy.

Chromosomal abnormality in the fetus

It has been estimated that more than half of all miscarriages occur because there is something wrong with the fetus.

Chromosomal abnormalities that result in severe deformities of the fetus will nearly always mean that the pregnancy ends in miscarriage, usually in the first few weeks. However, if the abnormalities are not so severe, the chances are that the pregnancy may proceed normally. For example it is thought that about a quarter of Down's syndrome pregnancies reach full term.

Many chromosomal abnormalities are 'mistakes' that occur during the development of the sperm or the egg – possible causes include the effects of certain drugs or chemicals or the age of the mother. In these cases there is often no reason why the woman should miscarry in a future pregnancy. However, in a few cases one parent may have a chromosomal abnormality that does not affect their health but which may produce a lethal chromosomal imbalance in the fetus. This can lead to recurrent miscarriage and therefore chromosome tests should be part of the investigations carried out on couples who have had two or more miscarriages.

Other fetal abnormalities

Other fetal abnormalities that may lead to miscarriage include spina bifida in which the spinal cord does not develop properly, and also a condition called anencephaly in which the baby does not have a properly developed brain. It is not known why these anatomical abnormalities occur but there is some evidence that they may be due to deficiencies or imbalances in certain essential nutrients. Recent research has shown that vitamin and folic acid supplements given before conception to a mother who has had a spina bifida baby may increase her chances of having a normal baby.

Problems in the uterus

An incompetent cervix, which is too weak to hold the contents of the womb until the pregnancy reaches full term, is one cause of recurrent miscarriage. It is not clear what causes an incompetent cervix, but sometimes it may be due to trauma in a previous birth, or possibly to hormonal factors. Other

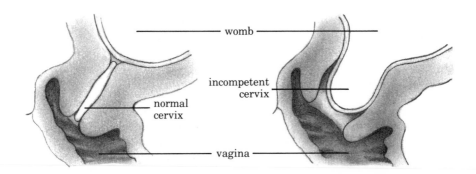

womb

incompetent cervix

normal cervix

vagina

women may just be born with a weak cervix. It is not always easy to diagnose an incompetent cervix but in general this diagnosis will be considered if there have been two or more miscarriages after the sixteenth week of pregnancy and there is no other obvious cause for them.

Treatment for an incompetent cervix may involve a cervical stitch or 'Shirodkar suture' as it is often known, in which stitches are inserted round the cervix to keep it closed during pregnancy. The stitches are removed when it is time for the baby to be born. Obstetricians vary greatly in how often they carry out this treatment for recurrent miscarriage, some considering that there is little evidence that it does any good at all. But the results of controlled trials should give more information soon. Meanwhile any woman recommended to have a cervical stitch should have the chance to discuss it with her consultant. The Stitch Network (see Useful Addresses), a branch of the Miscarriage Association, is also able to put women with this problem in touch with each other and to provide further information on the cervical stitch.

Some women have wombs that are an abnormal shape and this may prevent the fetus from developing normally so that miscarriage may occur, particularly in the middle three months of pregnancy. It is possible to correct some womb abnormalities with surgery.

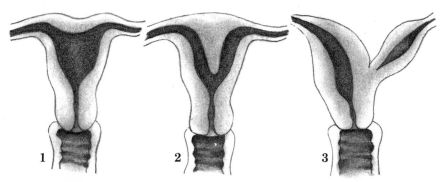

Uterine abnormalities
1 Slight abnormality indicated by the indentation along the top.
2 Partial non-union of ducts forming the uterus; the uterine cavity is divided by a ridge of tissue.
3 Severe malformation in which one uterine horn does not connect with the vagina. The accumulation of menstrual blood would require surgery.

Fibroids
Fibroids, which are non-malignant lumps in the uterus, are very common in all women whether they are pregnant or not. In some women they may cause heavy bleeding and cramps, and in a few cases they may distort the uterine cavity and produce miscarriage. However, many women experience

no problems and may have a normal pregnancy, unaware that they have fibroids. Troublesome fibroids may eventually disappear without treatment, but if necessary they can be treated surgically.

Infections and diseases

It is possible that severe fever, particularly in the early stages of pregnancy, may put the mother and the developing embryo at risk. Specific infections such as syphilis and brucellosis may lead to miscarriage but, although it is known that certain diseases can produce abnormalities in babies, it is difficult to establish a direct link between infections in pregnancy and miscarriage. Pregnant women are advised to avoid people with a serious infectious illness but minor infections such as colds and flu shouldn't pose any problems.

It is also possible that illnesses such as heart disease, kidney disease and diabetes may increase the risk of miscarriage. However, women with these conditions will always be watched by doctors from the early stages of pregnancy. Obese women are more likely both to have high blood-pressure and to be diabetic so they should do their best to get fit and lose weight before they get pregnant. They should also be carefully monitored during their pregnancy.

Immunological causes of miscarriage

Only half of the chromosomes of a developing fetus belong to the mother, so doctors have suggested that in some cases recurrent miscarriage may be due to the mother's body rejecting the fetus because half of it appears to be 'foreign'. It seems possible, therefore, that something goes wrong with the mechanism that normally protects the baby from this immunological reaction.

Research recently carried out has indicated that normally a pregnant woman produces antibodies against the white blood cells of the baby's father and somehow these are involved in protecting the baby from rejection by the mother. It has been found that some women who have suffered recurrent miscarriages do not produce these antibodies, but injections of their husband's or partner's white blood cells can give them a much better chance of a successful pregnancy.

It is clear that more research will have to be done on the immunological aspects of pregnancy, but it is already well known that a woman whose blood group is rhesus negative must be given an injection of anti-D gamma globulin soon after her first baby is born (or after a miscarriage or abortion) to prevent her rejecting and possibly losing a rhesus positive baby in her next pregnancy.

Hormonal problems

Hormones are very important in the maintenance of a normal pregnancy, so it is possible that a lack of certain hormones such as progesterone could produce miscarriage. Women are sometimes given hormones to treat recurrent miscarriage but so far results of trials testing the use of hormonal treatment to prevent miscarriage have not been very clear.

Other factors affecting the fetus

There are several other activities and environmental factors that can influence the success of a pregnancy. Alcohol, even a very moderate intake, can lead to bleeding and miscarriage. Although women who smoke are more likely to have a miscarriage, it is not known whether there is a direct link between smoking and miscarriage. However, smoking in pregnancy definitely produces smaller, less healthy babies so it seems likely that this activity will increase the likelihood of miscarriage.

X-rays, radiotherapy and certain drugs and poisons are known to cause miscarriage and, if their use is indicated during pregnancy, doctors will weigh up very carefully the necessity for them and the risk to mother and baby. No drug should be taken without a doctor's approval – it is probably a good idea to avoid painkillers such as aspirin, paracetamol and antibiotics unless absolutely necessary. And of course a woman should eat as healthily as possible, though not in excessive quantities, in the months before and during pregnancy.

Many women worry that sex during pregnancy may make miscarriage more likely. But so long as she is healthy and experiences no discomfort during sex, there is no reason for her to discontinue sexual activity. However, if she has had recurrent miscarriages, the doctor may recommend abstaining from intercourse particularly in the first three months of pregnancy. Vaginal examinations may also be restricted in women who have had miscarriages in the past as stimulation of the cervix may trigger premature labour. There is also a very small risk that other medical investigations such as amniocentesis, in which fluid surrounding the fetus is drawn out by means of a needle inserted through the abdomen, may cause miscarriage. For this reason these tests won't be carried out unless they are absolutely necessary.

Psychological problems

As with many medical problems, it does seem that factors such as stress and a woman's state of mind during pregnancy are related to miscarriage. It is already thought likely that stress and emotional upset can have a strong effect on menstrual cycles and on ovulation and fertility. And trials of new treatments for recurrent miscarriage often have a large placebo effect. This means that patients in a control group who do not know that they have *not* been given the treatment under test, but just receive the attention and care of medical staff, often have fewer miscarriages than similar patients who are not included in the trial at all. This indicates that simply feeling that someone is doing something to help and having some anxieties alleviated can increase the chances of a successful pregnancy in a woman who has had recurrent miscarriages.

It is difficult to prove that stress causes miscarriages since everyone has a different view of what is stressful. Up to a point a certain amount of stress is good for everyone. However, there is little doubt that a pregnant woman who experiences a very stressful event such as the death of a close relation is more likely to have problems. Unfortunately these stresses and worries are often difficult to avoid but practising yoga or meditation, reducing the demands of work – working part-time might be a possibility – and taking gentle exercise will all help to encourage a peaceful state of mind and complete relaxation during pregnancy.

It has been suggested that some women who have recurrent miscarriages may have a psychological problem that prevents the pregnancy reaching full term. Perhaps they are anxious about being a mother or were never really sure that they wanted a baby in the first place but just felt it was expected of them. Or they may have some other unresolved fear or anxiety that blocks successful pregnancy. Some reports indicate that counselling, relaxation or hypnosis may be able to resolve these hidden conflicts and increase the chances of producing a healthy infant.

What happens when a woman has a miscarriage

Every woman's experience of miscarriage will be different, but usually the first sign will be a coloured discharge or bleeding, although it is important to note that some women experience bleeding during pregnancy that is not serious and does not affect the baby. There may also be cramp-like pains or backache and a general feeling that all is not well. A woman should always see her doctor if she has these symptoms particularly if she has a history of miscarriage. She will probably be told to go to bed and rest, and if symptoms are mild she may continue to have a successful pregnancy and produce a healthy, full-term baby.

Another possibility is a 'missed abortion' in which the fetus fails to develop or dies. Symptoms of pregnancy will begin to disappear and miscarriage may occur at a later date, or the fetus may have to be removed either by dilatation and curettage (D and C) or by vacuum aspiration. However, if symptoms are more severe with increased bleeding or uterine contractions the woman will go on to have a complete miscarriage (or 'inevitable abortion').

Miscarriage itself can range from having an unusually heavy period at home to going through an experience similar to a normal delivery, depending to some extent on the reasons for miscarriage and on the stage of pregnancy. But doctors can do much to reduce any pain and discomfort that may be experienced. Another medical term, 'incomplete abortion', is used to describe a miscarriage in which not all the contents of the womb are expelled and will have to be removed by the doctor to eliminate the risk of severe infection.

Coping with miscarriage

Recovery from miscarriage will depend to a large extent on the support a woman gets from doctors, nurses and family. Feelings of emptiness and depression are common after miscarriage and women often feel guilty, thinking that something they have done, or their attitude to the pregnancy, has caused the loss of the baby. These feelings may be particularly strong if doctors are not able to give definite reasons for the miscarriage. Medical advances, including contraception, mean that women today expect to have great control over their fertility and the planning of their families. The result is that problems such as miscarriage or subfertility can come as a great shock, whereas in the past, when women often had many children and feared childbirth, miscarriage was sometimes a relief.

Doctors' attitudes

The attitude of doctors may not help very much either, particularly their use of the very emotive word 'abortion' to describe miscarriage in patients' notes, and their tendency to see miscarriage as a failure. Doctors and nurses, and friends too, will often react by saying something like, 'Never mind, you can always try again soon.'

This approach may work for some women – a successful pregnancy is the best cure in some cases, especially if it's the woman's first miscarriage. However, this may not be the best solution for women who find it difficult to accept the loss of the baby – they may have miscarried in the later stages of pregnancy – or for women who have trouble getting pregnant again. Being told to forget all about it may be the worst advice, opening the door to depression and more serious problems in the future. The miscarriage itself is just the beginning of the trauma as a woman will soon have to cope with the day when the lost baby was expected to arrive, with friends' children and new babies, with babies' clothes in shops. Everything will seem to evoke memories of the terrible loss, a loss which everyone around them seems to be trying to avoid talking about.

Talking about miscarriage

Women who have trouble coming to terms with the loss of a baby should be given plenty of opportunity to grieve over the loss, to talk about it and to express their fears about why it happened and their anxieties about future pregnancies. Unfortunately, hospital staff and GPs often do not have the resources or expertise to give enough support at this difficult time, but it is important to try to get as much help and information as possible from them, particularly about the possible reasons for the miscarriage. Writing down experiences of miscarriage and feelings, perhaps in a letter to a friend, can sometimes help to resolve the grief. In some cases a woman may need as much support as someone who has had a stillbirth or whose child has died, and the assistance of a professional counsellor may be needed to help a couple mourn the loss and to adjust themselves to the new situation. Some hospitals will allow parents to see the dead fetus, and sometimes this can help parents come to term with the loss. It also helps to be told that although people don't talk about miscarriage very much it is a common event and that most women who have a miscarriage will go on to have a normal baby. And also on the positive side, many couples find that going through an experience such as miscarriage means that they take less for granted in the future, and this helps them to improve their relationship with each other and their enjoyment of life, and to appreciate children they go on to have.

Getting pregnant again

It is usually considered safe to start having sexual intercourse again when bleeding has stopped – and if a woman does not want to get pregnant again immediately she should use adequate contraception straightaway. However, some women find that they don't feel like sex after miscarriage. There are many possible reasons for this. Hormone levels will take a while to get back to normal, she may still be upset and sex may be too closely connected with her experiences of the last few weeks – and with babies. But these problems should sort themselves out with sympathetic support from partner and friends.

Doctors may advise waiting between three and six months before trying to get pregnant again, though it is not unusual for a woman to conceive as early as one month after miscarriage. However, every woman is different and factors such as her age and the reason for the miscarriage will influence how long she waits before trying to get pregnant again. So long as a woman is physically fit there is no reason why she shouldn't get pregnant as soon as possible, but it is very important that she is not still too upset about the loss of the baby. The more physically and mentally fit a couple are at the beginning of the next pregnancy, the greater the chances that it will be a successful one. Keeping busy, possibly getting back to work as soon as possible rather than sitting at home waiting to get pregnant again, and taking lots of exercise will also help to get over feelings of grief and anger, and are good preparation for another pregnancy.

However, if a woman finds it difficult to get pregnant again or she miscarries repeatedly for no apparent reason, it is very important for her to keep the whole subject of pregnancy and children in perspective. If anxieties about getting pregnant or repeated miscarriages are putting chronic stresses and strains on partners and other children, it might be worthwhile for the woman to take stock of the situation and decide to put off trying again for a while, or even to stop trying completely and concentrate on those already needing her care and attention, particularly if she has other young children.

9 *Subfertility*

by Jenny Bryan

Once a couple have decided to have a baby they want, and expect, to get pregnant straightaway. After years of careful contraception with the fear that a single slip up could lead to disaster, many couples feel rather let down when they do not conceive during the very first month of unprotected sex. Were all those condoms and all that pill-taking really necessary, they ask themselves.

The fact is that it takes the average couple about five months of regular sexual intercourse without contraception to become pregnant. And one of the main determinants of how quickly conception occurs is the woman's age. Women are at their most fertile in their mid teens. Thus, a fifteen-year-old girl who has sex at around the time of ovulation has an eighty per cent chance of getting pregnant within the first month. This drops to fifty per cent by the time she is eighteen, twenty-five per cent at age twenty-six and just ten per cent when she reaches her mid thirties.

Most women are advised to wait for three months after coming off the Pill before even attempting to get pregnant. This is to enable the body to re-establish a normal cycle. And, while some women start having periods as soon as they stop taking the Pill, others can wait up to a year for their bodies to regain their normal monthly cycles, with the accompanying return to fertility.

It may sound obvious but getting pregnant also depends on having sex at the most fertile time of the month. According to the textbooks, a woman's menstrual cycle lasts twenty-eight days, with ovulation occurring about halfway through, on day fourteen. However, a large proportion of women do not conform to the rules; they may have short cycles and ovulate earlier in the month or have longer cycles and ovulate late. If it is going to occur, fertilization generally takes place within twenty-four hours of the egg being released from the ovary. Fortunately, male sperm live rather longer than eggs – four or five days. And they can lie in wait for the egg at the top end of the fallopian tube. So it is not necessary to time sexual intercourse to the minute. As long as it happens within two or three days before ovulation or up to twenty-four hours afterwards there is a good chance of conception. Even so, the difficulty of predicting ovulation means there is a large element of trial and error.

When to seek help

For a variety of reasons, therefore, even couples who ultimately prove their fertility with no help from the medical profession should not assume they will get pregnant immediately. It often takes up to a year and currently couples are advised only to seek medical help if there is no pregnancy after eighteen months to two years of regular sexual intercourse without contraception.

There are exceptions to this. Women over thirty, for example, may be advised to see a doctor after about a year. They are less fertile to start with and have less time to get things sorted out if there is a problem. Women of all ages who do not have regular periods should also seek help more quickly since they may not be ovulating often enough to conceive. And women whose periods suddenly become painful or who find sexual intercourse painful should also seek help since it is possible that they could have a pelvic infection in need of treatment.

It is quite understandable that people do seek help before they have given nature a chance. It is now well known that NHS help for those with fertility problems has not kept up with demand and people routinely have to wait months, sometimes years, to see a specialist, so they want to join the queue at the first sign of difficulty. The first port of call is the family GP, who can then refer a couple on to a specialist clinic. The GP can be a useful source of initial help and reassurance by checking that the couple have done all they can to help themselves. Either they may have mis-understood when the woman is likely to be most fertile or they may be

Causes of subfertility

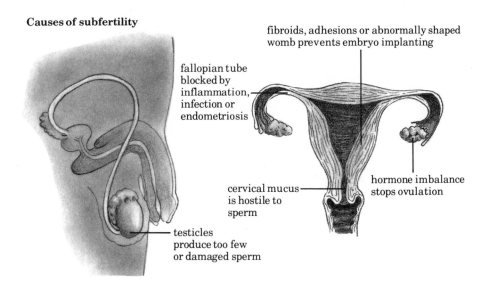

fibroids, adhesions or abnormally shaped womb prevents embryo implanting

fallopian tube blocked by inflammation, infection or endometriosis

hormone imbalance stops ovulation

cervical mucus is hostile to sperm

testicles produce too few or damaged sperm

experiencing difficulties with intercourse. The GP will probably also arrange for a sperm count for the male partner and advise the woman about using temperature charts to get some idea of whether or not she is ovulating. After that, it will be over to the experts.

Around one in eight couples find it hard to conceive and eventually find their way to a specialist clinic to begin the lengthy series of tests needed to find out what is wrong. Early on, all couples are urged not to blame each other especially when tests show who has the fertility problem. Neither partner is 'at fault.' In fact, fertility problems occur roughly equally in men and women. Treatment is being improved all the time but, sadly, less can be done to help men than women.

Causes of female subfertility

In order to draw a distinction between treatable and non-treatable fertility problems, it is now usual to apply the terms 'subfertility' and 'infertility' respectively. The reasons why women have problems in getting pregnant fall into four main categories:

- Failure to ovulate, generally caused by hormonal imbalance
- Blocked fallopian tubes, resulting from chronic inflammation, previous infection or surgery
- Failure of the embryo to implant in the womb perhaps because of fibroids, adhesions or abnormal shape of the womb
- Cervical problems, often due to the cervical mucus being 'hostile' to the sperm

Early on, there may be some clues to the cause of a woman's subfertility. Perhaps she has always had irregular periods or she has already received treatment for a pelvic infection. However, sometimes there may be more than one contributory factor, or both partners may have a problem. This is why it is so important for both to have tests and, once the possible cause is established, to understand exactly what the treatment is trying to do.

Testing for ovulatory problems

Even if there are no obvious signs of ovulatory malfunction it is usual to perform a series of hormone tests as the first step to assessing a woman's fertility. To understand what tests are done and why, it is important to understand just how it is that once a month, from menarche to menopause, we come to release an egg, or ovum. Every woman is born with all the eggs she will release during her reproductive lifetime – and an additional million or so she won't need! The control switch that triggers their release is in a gland at the base of the brain called the hypothalamus. At puberty this gland starts to release chemicals, appropriately called releasing hormones, which act on a second gland, close by, called the pituitary.

At the start of each menstrual cycle the pituitary releases a substance called follicle-stimulating hormone (FSH) into the blood-stream. This travels to the ovaries and stimulates growth of one or more eggs inside the ovaries. It is then a race between the eggs, each lying inside an outer coating or follicle, to see which one develops first. There can be only one winner and the 'also rans' simply disintegrate. While the egg race is going on, cells in the ovaries start to produce the female sex hormone, oestrogen. This acts on the womb, getting it ready to receive an embryo should the need arise. When the level of oestrogen reaches its peak it also has the effect of slowing down FSH production in the pituitary and triggering the hypothalamus to release the second of its releasing factors. Called luteinizing hormone-releasing hormone (LHRH) it in turn stimulates the pituitary to release a second hormone, called luteinizing hormone (LH). It is this hormone that, once it finds its way to the ovaries, triggers the release of the winning egg.

At around this time the ovaries stop producing oestrogen and start making a second hormone, called progesterone. This stimulates blood vessels in the womb and continues to make the womb receptive to any embryo that might chance by. If the egg remains unfertilized and conception fails to occur, levels of progesterone fall, the blood vessels contract and reduce their supply to the lining of the womb. Starved of oxygen, the lining starts to break up and fall away. As it does this the broken blood vessels bleed and the normal menstrual period begins.

Since each hormone has a role – either in egg release or preparation of the womb – any breakdown or blockage of the normal sequence of events can result in female subfertility. This is why FSH, LH, oestrogen and progesterone levels are the first to be measured to check for ovulatory function.

Home testing for ovulation

While these tests are being done, during the course of one or more cycles, a woman can do some simple tests for ovulation at home herself. At around ovulation there is a small but sharp rise in the body temperature (from about 36.4 up to 37 degrees centigrade) and if she takes her temperature religiously every morning it is possible for a woman to see if and when she ovulates. However, some women ovulate without any noticeable change in temperature, which is why the hormone tests are so important.

Recently, some manufacturers have introduced simple hormone tests that can be done at home to test for ovulation. These measure levels of LH in urine samples taken for six or seven days leading up to the likely ovulation day. Although more accurate, these are likely to have the same drawback as the temperature charts in that they will not recognize those women who ovulate without a significant LH surge. This is why specialist subfertility clinics run a whole battery of hormone tests to cover all eventualities. Many women attending a clinic are now also given an ultrasound examination at some point to check whether their ovaries look normal and are functioning properly.

Hormone treatment for ovulatory problems

If tests show that there is a hormone abnormality – and two thirds of ovulation problems are caused by this – the first priority is to correct it. The most commonly used drug is a product called clomiphene (Clomid). This boosts the pituitary gland to produce FSH and thus to start the whole process of egg maturation in the ovary. Clomiphene is therefore usually taken in the first four or five days of the menstrual cycle. It is usual to continue a course of clomiphene for at least six months and to monitor hormone levels at the same time to check that ovulation is indeed occurring.

Some women manage to get their eggs to mature in response to clomiphene but still fail at the last stage when the egg needs to burst out of its follicle and into the fallopian tube. For these women an injection of the pregnancy hormone, human chorionic gonadotrophin (HCG), at the moment when the egg is judged ripe for release can have the same effect as LH in triggering ovulation.

If these drugs fail, the next step is to try the so-called 'fertility drug', human menopausal gonadotrophin (HMG), marketed as Pergonal. This contains a mixture of FSH and LH and can be very effective in stimulating egg maturation and ovulation – sometimes too effective since there is a risk of multiple births when using this drug. It is given as a series of intramuscular injections in the first few days of the menstrual cycle. Alternatively, women may be given single hormone treatment with FSH alone to stimulate egg maturation, but this is even more expensive than Pergonal so tends to be kept as a last resort.

Recently, a further alternative has become available. This is to inject doses of the releasing hormones normally produced by the hypothalamus. These then trigger the pituitary to produce FSH and LH for itself and thus begin the natural chain of events. Doctors quickly realized that for this treatment to be effective the way in which the releasing hormones were given had to mimic exactly the normal activity of the hypothalamus which releases its hormones in pulses. In order to copy the hypothalamus, scientists have developed a pump about the size of a cigarette packet which pulses the drug through a needle implanted in the arm at the required intervals until ovulation has occurred.

In a small number of women it is not the hormones controlling egg development and release that are responsible for failure to ovulate. Instead, the pituitary produces too much of the milk-stimulating hormone, prolactin.

The body, seeing high levels of prolactin, assumes the woman is breast-feeding and does not need to ovulate. This problem can be overcome by using a drug called bromocriptine over a period of several months to suppress the prolactin levels.

These are the mainstays of treatment for women whose hormone tests show that they are failing to ovulate satisfactorily. If they are ovulating but the fertilized egg is failing to implant, progesterone injections or suppositories may correct this. All in all, hormonal abnormalities are about the easiest cause of subfertility to treat. Modern drugs, hormone replacement and the more accurate ways of giving them have enabled doctors to achieve a success rate of around ninety per cent in such women.

Damaged fallopian tubes

Roughly the same number of women with subfertility problems suffer from damaged fallopian tubes as fail to ovulate, and the number seems to be increasing. It is not an all or nothing phenomenon; your tubes are not either crystal clear or blocked beyond repair. There are many stages in between and in addition to some degree of blockage there are also women whose tubes are stuck to surrounding tissues. The lining of the tubes is highly sensitive and easily damaged yet it is crucial in feeding the egg and wafting it down to the womb. So, if for some reason it has become damaged, this too can lead to subfertility.

There are a number of causes of diseased fallopian tubes. Inflammation, with or without infection, is one of the most common causes of damage. This does not necessarily mean a venereal infection. An abdominal infection can spread to the fallopian tubes as can an infection following a miscarriage, an abortion or childbirth.

Another possible cause of infection is intrauterine devices. Some women find these painful and persistent excessive bleeding may signal that the IUD is causing inflammation. A further cause of tube damage is ectopic pregnancy. In a small proportion of pregnancies the embryo implants in the fallopian tube instead of the womb. This can be very dangerous if not recognized since the growing embryo may eventually rupture the tube or, at the very least, damage the lining and perhaps block the tube.

Some women's tubes become blocked or damaged because of a condition called endometriosis. The endometrium is actually the lining of the womb but in some women the cells that make up this lining are overactive and multiply too fast. Initially, this may mean that the uterine tissues are thickened, which can lead to heavy or painful periods when the lining comes away each month. But sometimes the overactive cells can spread outside the womb and start growing in and around the fallopian tubes, causing adhesions and blockages.

The fallopian tubes are thus very vulnerable to damage which, if left, can lead to later blockage and subfertility. This is why it is so important to recognize the first signs of infection – sudden pain in the lower abdomen, tenderness or raised temperature for example – and treat it before it has a chance to do any damage. Endometriosis can also be treated quite simply. In its mildest form nothing need be done. However, if there is evidence that it could damage the fallopian tubes, hormone treatment can be given to stop the endometrial cells from getting out of control.

How blocked tubes are diagnosed

X-rays are the standard procedure for diagnosing blocked fallopian tubes. To improve the picture of the uterus and tubes a dye is injected into the uterus through the cervix and allowed to find its way up into the tubes. If there is either a partial or complete blockage of one or both tubes this will show up on the X-ray as the dye will not be able to get through. This test is called a hysterosalpingogram.

If it looks as though there could be some kind of blockage, doctors generally like to take a closer look – to try to identify exactly what is wrong. This is especially true if the damage seems to be on the outside. The tube may be stuck to other tissues, such as the womb or ovaries, via adhesions. Adhesions occur as a result of inflammation. The body recognizes that something is wrong and cells grow to cover up and repair the damage. Unfortunately, this has the opposite effect because the extra cells simply gum up the works and cause the surrounding tissues to stick together.

The technique that is used for a closer look is called laparoscopy which is perhaps more familiarly used in female sterilization. A small incision is made in the abdomen, under general anaesthetic, and a mini-telescope-like tube is passed down into the abdomen alongside the fallopian tube. Not only can the doctor see the damaged fallopian tube, he or she may be able to effect minor repairs there and then by separating adhesions, for example, with tiny operating instruments passed down the laparoscope.

If there is more extensive damage a second operation will be required, this time through a larger incision. If the adhesions are limited to the outside of the tube it may be possible to separate them and free the tube so that it can work normally. Sometimes the top end next to the ovary gets blocked following infection and it may be necessary to free that. Alternatively, if it is the lower end of the tube that is blocked it may be necessary actually to remove the damaged part and rejoin the remaining ends, thus shortening the tube.

Microsurgical techniques
Crucial to the successful repair of damaged fallopian tubes during the last

ten years has been the development of microsurgical techniques. Surgeons now operate with the help of binocular microscopes that magnify the delicate structures on which they work many times over. Tiny operating instruments which look more suited to a doll's house enable them to separate adhesions without damaging the surrounding tissues.

The success of surgery in unblocking or otherwise repairing fallopian tubes varies enormously, according to the extent of the damage. And repairing tubes does not guarantee a pregnancy. All too often, scarring inside the tubes is sufficient to prevent conception even when the blockage has been cleared. Nevertheless, doctors tend to recommend that women have tubal surgery when their fallopian tubes are blocked before they consider trying for a test-tube baby. There may be a less than fifty per cent chance of a successful pregnancy following tubal surgery but even the most skilled doctors boast only a fifteen to twenty per cent birth rate for a single attempt at in vitro fertilization.

Sterilization reversal

Advances in microsurgery have also enabled doctors to repair a different type of tube damage, namely, sterilization. The success of an operation depends on what type of sterilization was done and on the amount of damage done to the tubes at the original operation.

Because of the uncertainties about sterilization reversal women are always advised to think very carefully before choosing this method of contraception and to consider it as permanent, except in exceptional and unforeseen circumstances.

Test-tube babies

Over 2,500 children have been born following in vitro fertilization since the birth of Louise Brown, the world's first test-tube baby, in 1978. During that time the technique has been polished and perfected so that pregnancy rates have risen from less than ten per cent to up to thirty per cent. The basic principles remain the same.

A woman who is accepted into one of the country's test-tube baby programmes must first undergo a series of hormone tests to see how often and when she is ovulating. Her partner too must have tests to check the quality and quantity of his sperm. The woman will then be given hormone drugs, such as clomiphene, to boost her natural egg development. This will be closely monitored both through hormone tests and through ultrasound scans of the eggs as they grow within the ovaries. It would be hoped that as a result of the hormone treatment a woman would produce three or four fully developed eggs instead of the usual one.

At the point when these eggs would normally be released from the ovary, doctors perform a laparoscopy to remove them from the ovary. A small incision is made in the abdomen and a tube carefully lowered into the ovary and each egg sucked up in turn. Doctors call it 'harvesting' the eggs. These are then carefully placed in a small plastic dish containing all the nutrients that painstaking research in the last ten years has demonstrated make up the optimal environment for growing embryos. Also in the dish go the precious sperm for the process of fertilization.

A single millilitre of semen from a normally fertile male contains anything up to 100 million sperm. But all too often it is found that women with a fertility problem also have partners with rather low sperm counts. This is why, for them, in vitro fertilization may be the most favourable option since in the laboratory far fewer sperm are needed for fertilization than nature requires in the fallopian tube. The fertilized eggs start to divide. After one to two days when each pre-embryo is just four cells in size, they are examined to check that they are normal and then returned to the womb in another small operation. Generally, the tiny embryos are gently pushed up through the cervix into the womb.

The next few days are critical. Additional hormone treatment may be given to ensure that the womb is as receptive as possible to the fertilized eggs. Research has shown that if three or four embryos are replaced this improves the chances of achieving at least one pregnancy. It does of course increase the likelihood of a multiple pregnancy if all the embryos take but most people feel that this is a risk worth taking.

A new variation on in vitro fertilization is a technique christened GIFT – gamete intrafallopian transfer. Here, the eggs and sperm are collected as before and mixed together but immediately returned to the woman's body, this time to the fallopian tube instead of the womb. The actual fertilization occurs inside the woman and not in the laboratory. GIFT is simpler and quicker than in vitro fertilization and may prove useful in couples where the man's sperm have difficulty swimming up through the cervix for fertilization.

Where is in vitro fertilization available?

Unfortunately, demand for in vitro fertilization far exceeds the availability of NHS facilities. There are a handful of clinics around the country that offer the service on the NHS but these have long waiting lists. A number of others rely on research funds and voluntary donations to keep their service going. The alternative is private health treatment. There are now a number of private clinics offering test-tube baby techniques at costs ranging from £1,200 on a daycare basis to over £2,000 for inpatient treatment. Some centres offer cheaper rates for subsequent attempts if women fail to get pregnant at the first try – which frequently happens, of course.

Uterine problems

Less common than either failure to ovulate or tubal blockage are problems in the womb itself. Women with such problems have little or no trouble in conveiving but the fertilized egg either fails to implant or latches on to the wall of the womb and later miscarries.

The most likely reason for this is growths on the inside surface of the womb. These may be small and nodular in which case they are called polyps. Or they may be much larger and make the womb swollen and misshapen. These are called fibroids. Neither of these types of growth is dangerous; many fertile women have them but it seems that in some cases they can prevent a fertilized egg from implanting. Both polyps and fibroids can be removed in simple operations; in fact, polyps can be scraped off the womb in a routine dilatation and curettage (D and C). This means dilating the cervix under general anaesthetic and passing a small instrument into the womb to remove the polyps. Removing fibroids is a rather larger operation since it means making an incision in the abdomen and opening the womb to remove the growths.

Some women are born with unusually shaped wombs – asymmetric or missing a piece at top or bottom – and this can often result in infertility. Corrective surgery is sometimes possible but there are no guarantees that this will enable a woman to carry a child to term.

A handful of women are born without a womb or with a womb so small that it cannot carry a child and there is not much that can be done in these cases. Scientists are working on ways of developing artificial wombs so that babies could develop, at least to some degree, in the laboratory. But this is something for the future. In the meantime, the only alternatives for women who, for one reason or another, cannot carry a child in their womb is adoption or the controversial business of surrogacy. In this latter case it would mean taking one or more eggs from the infertile woman, fertilizing them in the laboratory with her partner's sperm and then replacing the pre-embryo in the womb of the surrogate mother who would carry it through pregnancy and give birth.

The hostile cervix

In addition to playing a vital part in the first stage of labour the cervix has the rather more mundane job of producing mucus. At around the time of ovulation the cervix produces approximately twenty times the usual amount of mucus. Its texture also changes; it becomes clear and watery and it is at this time that sperm find it easiest to swim up through it into the womb. This makes sense as it is at this stage of the menstrual cycle that the woman is at her most fertile.

Various abnormalities of the cervical mucus and/or the sperm can lead to fertility problems. If the mucus remains thick during ovulation the sperm find it hard to swim through it to fertilize the egg. If there is insufficient mucus this can also impede the sperm's progress up through the womb. More recently, scientists have found that some women have the right quantity and quality of mucus but it still does not permit some sperm to pass through. This is thought to be an immune incompatibility; the woman's cells recognize her partner's sperm as foreign and refuse to allow them through. This can be very specific; one man's sperm are refused entry while another's allowed through.

The sperm themselves can add to the problem. It is not just a question of quantity but also of quality. All men produce a proportion of misshapen or immobile sperm which float helplessly around the vagina instead of heading for the fallopian tubes. But if too large a proportion of the sperm are inactive or deformed it becomes difficult for fertilization to take place. If in addition to this the woman's mucus is 'hostile' it becomes almost impossible to get sufficient sperm to the egg.

Tests can be done on both the mucus and the sperm to identify this form of fertility problem. It is also necessary to get a sample of postcoital mucus to see how the sperm are getting on. So this means getting to a subfertility clinic soon after sexual intercourse has taken place. While identifying that there is a hostility problem it is less easy to correct it. The most successful technique is in vitro fertilization or GIFT. This bypasses the need for the sperm to swim through the mucus and also enables scientists to pick the healthiest looking sperm and the strongest swimmers.

When all the tests are negative

By listing all the main causes of female subfertility in this way, it may seem as though the procedures for diagnosing and treating these problems are quite straightforward and done in sequence. They are not. Many of the tests are done alongside each other. So doctors will probably take some X-rays and ultrasound scans of the ovaries, fallopian tubes and womb to see if there are any obvious anatomical abnormalities at the same time as performing hormone tests.

Results may be equivocal or borderline. On one occasion the results are normal and the next they are not. Hormone levels may be found that would be sufficient in most women but still the patient does not get pregnant. Or the hormone levels would not on their own be a problem but, coupled with a tubal problem or low sperm count, tip the balance towards infertility.

Then there are the couples whose tests are all negative; both partners seem perfectly healthy yet they do not conceive. All the doctors can do is to advise them to go away and keep trying. Most people who work in sub-

fertility clinics believe that there is a psychological component to subfertility. They often see a couple in the depths of despair because they are
unable to conceive. Then, when finally the pressure is off and they have
accepted their childless state or are considering adoption, the woman gets
pregnant.

Artificial insemination

When the male partner is infertile some couples decide to try artificial
insemination by donor (AID). This means that a sample of semen from an
anonymous donor is injected through the woman's cervix at the time of
ovulation. AID services are few and far between as are specialist subfertility
services in some parts of Britain. The British Pregnancy Advisory Service
(BPAS) does offer AID, at a cost that most couples can afford. Some
subfertility clinics run AID services and others will know where it can be
done locally.

It is still possible to adopt

Most people are aware that the supply of babies for adoption has all but
dried up with the introduction of effective contraception and easily available
abortion. However, there are still children available for adoption.

Most of the children who are adopted nowadays are older, of mixed race,
socially disadvantaged or handicapped. The British Agencies for Adoption
and Fostering produces a book called *Be My Parent* which describes 150
such children. And recently it has started producing video cassettes of
some of the children. Any qualms people had about the advertising of
children in this way have largely disappeared following the success of the
scheme. Some infertile couples may not find it possible to adopt such
children but for others it is the answer to years of misery trying and failing
to have a much-wanted family.

For those still trying the various subfertility treatments, and those
trying to come to terms with the fact that they will not have children, there
are self-help and support groups. The best known of these is the National
Association for the Childless which can put people with fertility problems
in touch with other couples in a similar situation.

Talking to others who cannot conceive does not take away the heartache
but it can help subfertile couples to realize that they are not alone and they
can talk about the various investigations and treatments they are trying.
And if these fail they can see how other people can eventually make happy
and contented lives for themselves without children.

10 *The Menopause*

by Jenny Bryan

The menopause – or change of life, as it is often called – is the time after which a woman can no longer have children. It does not happen overnight; it can take months or even years. It occurs most commonly between the ages of forty-five and fifty-five with an average age of fifty-one. But it can happen earlier or later than that. There are many unverified stories of women giving birth well into their seventies. More reliable is the case of an American woman, Ruth Kistler, who gave birth to a daughter in 1956 when she was fifty-seven and a half years old. And there are a number of women this century who gave birth in their mid fifties.

Although a woman's fertility declines sharply during her late thirties and forties, it is because her periods do not stop immediately at the menopause that she is advised initially to continue with contraception because of the small risk of pregnancy. Generally, the advice is to continue contraception for two years after the last period under the age of fifty and for one year if the menopause occurs over fifty.

During the menopause a woman's ovaries gradually stop producing oestrogen, her egg production and periods are erratic and eventually they stop. The pituitary gland in her brain, sensing the fall in oestrogen levels, attempts to compensate by pumping out large amounts of the hormones follicle-stimulating hormone (FSH) and luteinizing hormone (LH) which normally push the ovaries into action – but to no avail. However, the high levels of these hormones are useful in indicating to the doctor that a woman is indeed starting the menopause – and is not pregnant – when she reports a missed period or two.

What else happens at the menopause?

There are dozens of myths surrounding the menopause: you'll go potty, you'll start shoplifting, you'll no longer want to make love, you'll get fat and grow a beard! What nonsense! About a quarter of women sail through the menopause, scarcely aware that anything is happening at all – apart from the obvious disappearance of their periods. Some fifty per cent do experience one or more menopausal symptoms, such as heavy or erratic periods, hot flushes, vaginal dryness or depression. They are fairly minimal, lasting perhaps up to a year, and maybe requiring the occasional visit to the doctor. The remaining twenty-five per cent of women do experience symptoms that are sufficiently severe to make the menopause a difficult time and these require effective treatment. It is important for women not to attribute clinical symptoms to the menopause without consulting a doctor, since they may be due to an unrelated illness.

The disappearance of oestrogen is at the root of most menopausal symptoms. Oestrogen, in addition to its role in preparing the womb for a baby, has many other functions throughout the body. So directly, or indirectly, it is involved in hot flushes, night sweats, vaginal and period problems and, later on, in thinning of the bones – a condition called osteoporosis.

Menstrual problems
Virtually all women going through the menopause will experience some disruption of their menstrual cycle before their periods stop altogether. Just as it takes several years for her periods to become regular after a girl starts menstruating so it takes some time for them to wind down and stop. This means that sometimes a period will be light, sometimes heavy. There may be some bleeding mid cycle and it may be rather difficult to predict when a period is going to occur even in the most 'regular' women. Needless to say, this is all very trying to women who have, for upwards of thirty years, been used to taking their periods in their stride and not allowing them to interfere with their daily routine. Suddenly, they are never quite sure when a period may start and when it will end.

Hot flushes
About seventy per cent of women suffer from hot flushes during the menopause and they are generally accompanied by night sweats. A flush can last anything up to five minutes and usually occurs between ten and twenty times a day. A woman feels hot all over, not just on her face and neck as if she were blushing, and there is rarely any warning that it is going to happen. She sweats profusely and then may start shivering. Night sweats follow a similar pattern except that, because she is in bed, the woman is woken to find herself, and often the bedclothes, drenched in perspiration.

Painful intercourse
Women who complain that sexual intercourse is uncomfortable or painful are generally found to have a thinning of the lining to the vagina. This happens to a greater or lesser degree in all menopausal women as the lack of oestrogen makes the outer layers of the vagina fall away and the remaining tissue less elastic. In addition, she may no longer produce sufficient of the juices that lubricate the vagina, especially during inter-

course, and so she may feel dry and sore. Changes also occur outside the vagina. The folds of skin – labia – covering the opening to the vagina slowly shrink and the pubic hair is gradually lost.

Osteoporosis

Bone loss seems to occur because the fall in oestrogen levels interferes with calcium metabolism and calcium is an essential part of bone. The small amount of bone loss that occurs before the menopause is accentuated after oestrogen levels fall away and the bones become increasingly thin and fragile. It is generally in old age that this effect becomes really noticeable.

Above: An X-ray representation showing the hip-bone of a young woman (left) compared with that of an older woman with osteoporosis (right). Note how the bone of the older woman has thinned out, making it brittle and prone to fracture. *Left:* A woman with osteoporosis being supported by the Director of the Osteoporosis Society.

Many women actually shrink an inch or two and, in the worst cases, the spine and neck become so severely bent that the woman finds herself looking permanently down at her feet. Osteoporosis affects one in four women and one in two women over seventy break an arm or leg because the condition has made their bones so fragile.

Other symptoms

As well as these most common physical effects, the menopause produces a number of other less specific symptoms. Some women get palpitations, others feel dizzy or are prone to indigestion. A few get an itching sensation all over their skin as though an insect were walking over them. Sometimes the thinning of the tissues that occurs in the vagina also happens in the bladder so that there is some loss of control, resulting in a form of incontinence.

It is uncertain how much of a role the menopause has in the rest of the ageing process. Many women complain that their skin starts to sag and their hair thins and loses its shine. Those extra pounds put on at Christmas become much harder to lose. There seems to be an energy gap and, after the menopause, women find that they simply cannot manage some of the things they were doing ten years before. But the ageing process does not start suddenly at forty-five or fifty. Brain cells start dying almost from the moment we are born – never to be seen again! A woman of thirty has lines around her eyes and mouth which she didn't have at eighteen. Her face is stamped with experience. Her waistline is probably no longer that of a teenager, a few veins are apparent in the backs of her legs.

If all these signs of ageing could simply be attributed to hormones then it should be quite easy to find the elixir of youth. But they are not. No one understands all the reasons why we age; some of the signs may be more apparent after the menopause but many would probably have occurred with or without the loss of hormones. So it is no good blaming it all on the menopause.

In the longer term, women become more prone to heart disease after the menopause and they are more likely to require treatment for high blood-pressure. It seems that until the menopause their hormones help protect them from the various circulatory problems to which men are prone in their late thirties and forties. After the menopause, however, they gradually become as susceptible as men.

The psychological effects

It is hardly surprising that an event that marks the end of a woman's reproductive life has psychological as well as physical effects. At the very least there will be a certain sadness at the passing of an era accompanied and reinforced as it often is by children leaving home or the appearance of grandchildren. Just how strongly a woman feels these changes can depend on how satisfied she has been with her lot. If she has felt generally happy and contented with what life has brought her she will probably see the menopause as the start of something new; a time of fewer responsibilities and more time for herself and her partner or friends. But if life seems an endless struggle with few rewards and little security, the menopause may seem the final insult – even her womanhood has been taken away.

During the menopause even the sanest and most positive women need a little extra thought and consideration from those around them. They may well feel tired and below par – not least because their sleep has probably been disrupted by night sweats. They may feel anxious and need reassurance, or they may be depressed and need cheering up. It is usually the women who have massive changes of character – who shoplift or go on wild spending sprees, shout and scream, take to drink or disappear – who hit the headlines. But such cases are extremely rare. You aren't suddenly going to become a Jekyll and Hyde character!

Sexual relations
Many women worry that they will suddenly become totally unattractive to their husbands or partners and lose all interest in sex. These fears, too, are groundless. True, some women may not feel like making love if vaginal dryness is making it uncomfortable or painful. And if they are suffering from insomnia they may not feel up to a night of passion. In contrast, the menopause often brings a new lease of life to a couple's sex life. Once her periods have stopped there are no more worries about getting pregnant and no need to be organized about precautions. Therefore sex can be more spontaneous.

This does not mean that every couple must feel the need to prove that they are still young and sexy. Over the years many couples do reduce the frequency with which they have sex and this has little to do with the menopause. They simply feel less need and desire for sex and are quite happy in each other's company without frequent intimate physical contact. There is no need to feel ashamed or embarrassed about separate beds or separate rooms. The only important thing is that both partners are happy and content with how they have chosen to live together.

Don't suffer in silence

Many women do not like to go to the doctor about menopausal symptoms because they don't want to make a fuss. They know that the menopause is a natural event and believe that they must simply grin and bear it. That is not so. Fortunately, doctors are becoming more sympathetic to the problem and they can help with drugs for many of the physical symptoms and support and counselling for the psychological ones.

Before beginning any treatment the doctor will first want to be very sure that he or she is actually treating menopausal symptoms. There are other reasons for abnormal bleeding, tiredness, palpitations and so on, and the doctor will want to exclude these by physical examinations and tests. A growing number of hospitals now run menopause clinics and you can ask to be referred to one of these. Not only are the medical staff up-to-date with the latest thinking on all the latest treatments for the menopause, but you will also be able to meet other women with problems similar to your own and hear how they cope with them.

Hormone replacement therapy

Hormone replacement therapy (HRT) forms the basis of treatment for many of the physical problems that accompany the menopause. But it cannot cure them all. As the name suggests, HRT puts back the hormones – in particular the oestrogen – that are no longer produced after the menopause. It is therefore very effective in treating and preventing the oestrogen-related symptoms – hot flushes and sweats, vaginal dryness and discomfort, and osteoporosis.

HRT has had a chequered career. Hailed as a godsend in the mid 1950s when it was first introduced, HRT lost favour during the early 1970s when it was found that high doses of certain oestrogens could cause abnormal growth of the lining of the womb and, in some cases, cancer. However, towards the end of the decade it was discovered that smaller doses of oestrogen, given with intermittent progestogens, removed the risk of cancer and the only women who may still be advised not to have HRT are those who have had breast or womb cancer.

Many women ask why HRT is safe for menopausal women when doctors advise against the contraceptive pill for women in their forties because of the increased risk of circulatory problems. After all, both drugs contain the same hormones – oestrogen and progestogen. The fact is that the amount of the two types of hormone is much lower in HRT than it is in the Pill. And since menopausal women make much smaller amounts of these hormones than fertile younger women, their arteries are not exposed to dangerously high levels even when they are on HRT.

HRT can be given in several different ways. The most common and simplest method is to take the hormones in tablet form. This means daily oestrogen for three weeks supplemented with progestogen for ten to thirteen days each month. Oral treatment like this is geared to helping women with a whole range of oestrogen-related menopausal symptoms. Increasingly popular are hormone implants. These are pellets that are inserted under the skin and slowly release their hormonal contents over a period of three to six months. This has the advantage that there is no need to take tablets each day, but some people are nervous about using implants which, once in, cannot be removed.

Some women who are suffering from vaginal problems prefer to apply oestrogen creams. These are easily inserted into the vagina every day for a couple of weeks while the condition is being sorted out and then two or three times a week to prevent it recurring. If the problem is simply one of vaginal dryness, women are usually advised to try lubricating creams, such as KY jelly. These can be very effective in making up for the lack of cervical secretions after the menopause so that HRT may not be needed.

Basically, there are two schools of thought regarding HRT. Some doctors believe it should be reserved only for women with severe oestrogen-related menopausal symptoms, while others believe that all women should get it – at least for five to ten years after the menopause because of its effectiveness in preventing or delaying osteoporosis.

There is no question about the potentially disastrous effects of osteoporosis – one in ten women who break a hip through osteoporosis will die from it. But in spite of the fact that one in four women will get some degree of bone thinning it is difficult to predict who will be crippled by it. And if the odds are against disabling osteoporosis can you justify taking drugs for five or ten years? There is no evidence to suggest that current HRT has any long-term adverse effects, but there are still doubts in the backs of some people's minds about the effects of years of HRT on the womb. Perhaps more of a disadvantage to most menopausal women is the fact that by having HRT they continue to have periods. Menopausal symptoms, however severe, do pass, but few women would relish the idea of regular monthly bleeds well into their sixties and seventies.

What are the alternatives to HRT?

Whether or not women have HRT it is important that they take other simple steps to keep in good health and reduce the risk of osteoporosis. Diet is top of the list.

A balanced diet with sufficient protein, fresh fruit and vegetables, and not too much refined carbohydrate and fat, is recommended for anyone wishing to keep healthy. But it is especially important at times of stress, such as the menopause. Unfortunately, it is at just such times that diet may get forgotten. If the children have left home it may not seem necessary to cook regular meals. A series of snacks, biscuits and cups of coffee start to replace the meat and two vegetables, fruit and cheese. And it doesn't take long for a poor diet to show up in dry and lacklustre skin and hair, broken nails, tiredness and a general run-down feeling.

Since osteoporosis results from lack of calcium it makes sense not just to ensure a balanced diet but actually to boost calcium intake. Calcium is found in a variety of dairy products, including milk, cheese and yoghurt. People who are worried that these foods also contain large amounts of the saturated fats linked with heart disease can either buy low-fat dairy products or they can take calcium supplements. Some food manufacturers are already fortifying things like bread with calcium so it's worth checking on packaging. In general, though, the advice is to ensure that you are consuming at least 1,500mg of calcium per day.

Vitamin supplements

Regular exercise and plenty of fresh air are also important; exercise to keep bones and joints active and to keep weight down, and sunlight to make vitamin D which is important for strong, healthy bones. A lot of claims are made for other vitamins. Vitamin E, as well as its fabled youth-giving properties, is said by some to be very helpful in preventing and treating hot flushes and vaginal problems. Proponents of this treatment recommend that it should be accompanied by vitamin C. The vitamins can be taken either as capsules or vitamin E can be applied as a cream in the vagina. B6 is another vitamin of value to some women with menopausal symptoms. It is widely used, with or without evening primrose oil, to relieve pre-menstrual tension (see Chapter 4) and many women suffering from depression during the menopause believe that both these supplements help them to overcome this.

The evidence in favour of vitamins E, C and B6 to relieve menopausal symptoms is controversial. Some studies have shown improvements, others have not. If you decide to try vitamin supplements be sure that you get good advice, since vitamin E, for example, can be dangerous in large doses. Some doctors, though sceptical of their effectiveness, suggest that women try these treatments and see how they get on. If their hot flushes and vaginal discomfort seem to improve, they should stick with the vitamins. If there is no improvement within a few weeks, however, perhaps you should consider more conventional treatment such as HRT. The same goes for homeopathic and herbal remedies. Many women swear by these remedies and if they work for you, that's all that matters. But be sure that you go to reputable practitioners – the Institute for Complementary Medicine or the British Homeopathic Association (see Useful Addresses) keep lists of trained people in all areas of the country.

Sorting out the emotional problems

Ideally, it is best to get to the root of emotional problems rather than merely treat the symptoms. This is true whether or not they are related to the menopause. However, this may be easier said than done. Anxiety, insomnia and depression are common around the time of the menopause and are frequently interrelated. Sadness at the passing of the reproductive years, difficulty adjusting to children moving away, fear of becoming fat and unattractive, and worries about the future can all contribute to emotional problems at the menopause. And love, support and reassurance from family and friends can go some way to alleviating these causes of anxiety and depression.

It may be difficult for an eighteen-year-old, living away from home for the first time, perhaps away at college or starting a new and exciting job, to understand that mother needs a little extra consideration. At any other time the mother may laugh off the fact that her teenage children are using her home like a hotel and laundry service. But when she is suffering from hot flushes, feeling tired and unattractive and is unsure quite where she fits in any more, thoughtless words from husband or children may be enough to tip her over into tears and recriminations.

All too often this sets up a vicious circle: mum in tears again, must be her hormones, better leave her alone while she gets over it. Being left alone on her own serves only to reinforce her uncertainties about herself and the feeling that she is simply taken for granted – a part of the furniture. It would be far more helpful for mother to be able to discuss her worries and fears – however silly they may seem – with her family, so that they can understand the thoughts going through her mind and try to do something to alleviate the worries.

At the most simplistic level, she may need her husband to tell her that he still finds her attractive, to reassure her that she hasn't suddenly turned into a sexless drudge. Or her children can tell her that, although they may seem to be away leading full and exciting lives, they still need to share both the ups and downs with someone back home. Whether or not she is feeling tired and under the weather, she will surely appreciate some extra help around the house, a few evenings out, something to brighten up her wardrobe. But, most important of all, she needs understanding and consideration, not just a few mutterings about when the menopause is over and things can get back to normal.

How counselling can help
Some women may feel that they need to talk over their fears with someone outside the family. This may be because, however many hints they drop, their partner and children simply do not seem to want to know. Or they may feel their worries are too trivial or foolish to discuss with someone close to them and they prefer a more impartial listener. For them the answer may be counselling. Some GPs run counselling sessions while others believe that counselling should form part of every consultation and they have longer appointments to accommodate it. However, counselling is not something that all doctors are good at, so some practices employ a counsellor specially to help people with emotional problems.

In its simplest form, counselling is listening – giving someone the opportunity to voice their problems and to know that someone cares. Counsellors should not force their own views on their clients. Ideally, they should help people to identify their problems and the possible courses of action. The final decision about what to do should be that of the person being counselled, not the counsellor. If your doctor is not interested in counselling or you do not feel that you can approach him or her, you can contact the British Association for Counselling (see Useful Addresses) to find someone in your area who can help.

Coping with stress

For some women, stress is at the root of much of their anxiety; they may not feel that they are coping with life or they may feel that they are overreacting to relatively minor worries. Some benefit from relaxation or yoga classes to help them unwind and see their problems in a new and less worrying light. Most areas have such classes nowadays and some people find that relaxation tapes to play at home are also very helpful. Stress is not simply a trigger of emotional problems, it can have physical effects too. And women who suffer from palpitations or indigestion during the menopause may find relaxation classes helpful.

Unfortunately, some anxieties do not go away with counselling or relaxation. Very effective in these circumstances are short courses of tranquillizers to take away those daytime anxieties, and a few days or weeks of sleeping pills to help restore a normal sleeping pattern. But both types of drug should be seen only as a helping hand when ti .es are bad, not as a long-term prop. Tranquillizers and sleeping pills are addictive and it can be difficult to get off them after courses lasting more than a few months. The same goes for antidepressant drugs, but if the depression is severe drugs may be the only answer and they can be very effective.

Some sexual problems do have an emotional basis and may also be relieved by drugs – but tranquillizers or antidepressants are not the answer. If sex is painful lubricating gels or HRT may be very effective. But if the problem is one of libido other hormones can be very helpful. Women who complain that they simply do not feel like sex during the menopause – and want to do something about it – may be prescribed small amounts of the male hormone, testosterone. This will not turn them into men! But it can help restore their desire for sexual intercourse.

The menopause is not an illness or a disease – it is a natural event that happens eventually to all women who reach middle age. Women approaching the menopause should not expect months or years of physical and emotional problems. Whether they opt for rest and relaxation, vitamins and homeopathic remedies, hormone replacement therapy, short courses of sleeping pills or antidepressants, or a mixture of all of them, they should ensure that everything possible is being done to relieve their symptoms. If one thing does not work, try something else. The secret of a sane menopause is to persevere and not to suffer in silence!

11 *Hysterectomy*

by Antonia Rowlandson

A hysterectomy is an operation to remove the uterus, or womb, and possibly the ovaries as well. It is a major operation and many women find that it takes several months to feel completely fit again afterwards. However, provided the operation is done for good reasons, and the woman is well informed of the physical aspects of the operation, and well prepared emotionally, there is no reason why she shouldn't make a full recovery. In fact, many women find that the operation makes them feel much better, though this will depend on why the operation was carried out in the first place. Also there will be no more worries about periods, contraception or the possibility of developing cancer in the organs that are removed.

Hysterectomy is a very common operation – each week more than 1,000 women in the UK have this type of surgery. In some cases the operation must be carried out to treat a serious condition such as cancer. However, on many occasions the uterus is removed to alleviate less serious problems such as fibroids or menstrual difficulties.

A gynaecologist may recommend a hysterectomy to treat:

- Cancer of the uterus or ovaries
- Fibroids: areas of benign, fibrous tissue which may develop in the walls of the womb and can cause heavy bleeding and pain
- Endometriosis, in which material similar to that lining the womb grows in other parts of the pelvic area causing discomfort
- Menstrual problems, particularly heavy, irregular periods
- Prolapse of the womb, which means that the womb has dropped due to a weakening of the ligaments that support it
- Severe pelvic inflammatory disease

Is the operation really necessary?

There is little doubt that if a hysterectomy is recommended to treat cancer of the uterus or ovaries then it is probably a good idea to take the gynaecologist's advice and go ahead with the operation. However, the criteria used to decide whether to carry out a hysterectomy to treat less serious problems such as menstrual irregularities are less clear. For this reason it is very important for a woman to find out from the gynaecologist exactly why he is recommending the operation and how it can help solve her problems. Then she can be involved in making an informed decision as to whether the operation would be a good idea for her.

If there are any doubts about the necessity of the operation, it is very important to discuss them with the gynaecologist, GP and partner or other member of the family and friends. The opinion of a second gynaecologist can always be sought if necessary. A hysterectomy should not be rushed into when recommended to treat minor, non-life-threatening conditions, especially before all other alternative treatments have been considered. Hysterectomy may seem the easy way out to a woman (and her doctor!) if she has been suffering from distressing gynaecological problems for some time, but it is important to remember that it is a major operation and, like any operation, is not without it is own complications and side-effects. No woman should subject herself to major surgery unless there is a good reason for it and unless she can expect some improvement in her health as a result of the operation.

Possible alternatives to hysterectomy

Fibroids, endometriosis and menstrual problems may all be treated with drugs and both prolapse and fibroids may be successfully treated with less drastic surgery. Prolapse and related problems of incontinence may also respond to exercises that strengthen the perineal muscles – yoga is thought to be one of the best forms of exercise for toning up these muscles. It might be worth finding a yoga teacher who appreciates the use of this form of exercise for women's problems such as these.

It may also be worth considering that some external stress, such as worries about family, relationships or work, might be at the root of problems such as irregular periods or heavy bleeding. It is well known that emotions can affect hormone levels and so influence menstruation and fertility. So in some cases sympathetic help with these problems from a GP or professional counsellor may sort out the stresses and be a practical alternative to surgery. Hysterectomy may turn out to be the only effective treatment for severe cases, but these less drastic approaches should be tried first if possible, particularly if a woman is reluctant to have the operation.

A dilatation and curettage (D and C) may be carried out to clear up painful, heavy periods. This procedure is performed under general anaesthetic and involves dilating or stretching the cervix and gently scraping out the lining of the womb. It may also be used to remove the remnants of a pregnancy after childbirth or abortion, or to diagnose cancer. But a D and C may only improve heavy periods for a few months, after which the only other option may be a hysterectomy, in cases where drug treatment has been equally unsuccessful.

Preparing for the operation

Once a woman and her doctor have decided that a hysterectomy is the best way of dealing with her particular problem, she will probably have at least a few weeks to wait until the operation is actually carried out (unless the operation has to be done more urgently to treat spreading cancer). This should give her plenty of time to plan her working life and to make arrangements for the care of her house and her family while she is in hospital and when she is convalescing at home. In general, good preparation – both practical and emotional – means fewer problems and a speedier convalescence after the operation. Friends, and relations, including children, should be made aware that they can't expect a woman who has had a hysterectomy to return to her usual role within a few days of coming back from hospital.

Physical preparation

As with any operation, it is important to be as fit as possible when it is carried out. Eating well and taking lots of exercise in the weeks before the operation, and not being overweight at the time, will mean that a woman is more likely to make a speedy and successful recovery. It is also a good idea to stop smoking a few weeks before the operation as this can increase the risk of complications during and after surgery. And women using oral contraception will need to change to another method about a month before the operation because the Pill increases the chances of developing blood clots.

Emotional preparation

There are some women who feel tremendous relief when the gynaecologist recommends hysterectomy – perhaps they have suffered years of painful, heavy periods or uncomfortable sexual intercourse. However, the reaction of most women when they are told that they are to lose their womb is one of shock, particularly if they do not feel that they have reached the end of the childbearing period of their life.

Whatever anyone says, it is *not* the same as having your appendix or tonsils out. Even if a woman has had her family, the loss of the ability to produce children can seem to threaten her role in life as a woman and a mother. Feelings often expressed as the time of the operation draws near include: 'I'll be less of a woman' and 'Sex will never be the same again.'

Of course, losing the womb can be distressing especially if children were part of future plans. However, if the operation is carried out for good reasons which the woman can accept, she should try to come to terms with this loss. She should acknowledge that it *is* a loss in some respects and let herself be unhappy about this for a while ('mourning' the loss); then she should start concentrating on the positive aspects of the operation. It will improve her health, even perhaps save her life if carried out to treat cancer, and contraception and periods will be things of the past. She will certainly not be less feminine and there is usually no reason why she should have any problems with sex. In fact, many women find that their sex lives improve after hysterectomy, when their original problem is solved and contraceptive measures no longer interfere.

A woman who has both her ovaries removed as well as the womb before she has reached the menopause should be prepared for a slightly longer period of adjustment as hormonal levels will be upset, making her more vulnerable to depression and sexual difficulties. In these cases, extra support from family and doctors, together with hormone treatment, can usually sort out all these problems.

Partners' reactions to the operation

The reactions of a woman's partner to her hysterectomy operation play an important role. A woman needs all the support she can get before, during and after an operation and a man who has fears about the effect of the operation on his partner's health, and on their future relationship and sex life, may not be very helpful. A man may feel guilty, thinking that in some way he has caused his partner's problems. He may be worried about hurting the woman when sex can begin again a few weeks after the operation. Involving a partner in the decision to have the operation and, if possible, talking to him about what it involves and working out fears together should put him in a much better position to give valuable support. Talking to each other will of course be particularly important in cases where the couple were hoping to have children some time in the future.

Partners may be reluctant to talk about the operation and its consequences, in which case the help of the GP or district nurse may be needed to talk to a couple together. Talking to a friend whose wife has had a hysterectomy may also help. There are of course many women who have a hysterectomy who do not have a partner – they may be single, divorced or widowed. In these circumstances other relatives and friends, particularly those who have been through the same experiences, will be invaluable. Hysterectomy support and self-help groups may also be able to help.

The different types of hysterectomy operation

Most women have some fears about what the operation will actually involve and are worried about what parts the surgeon will remove and what he will leave behind. Knowing a little more about the operation can help to allay these fears, and all women should attempt to get an adequate

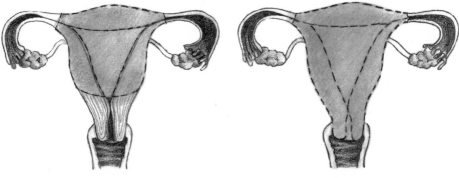

Subtotal hysterectomy   **Total hysterectomy**

explanation of what is involved from their gynaecologist or GP.

There are several different types of hysterectomy operation and the type of operation carried out will depend to a large extent on the initial reasons for surgery.

Subtotal hysterectomy
In this operation the womb only is removed – the ovaries and cervix are retained. The drawback of this operation is that it leaves the cervix behind as a possible site for the development of cancer in the future. So a subtotal hysterectomy is carried out only if a more extensive operation is more difficult or dangerous. However, a subtotal hysterectomy which does not remove the cervix is easier to perform and is also the hysterectomy operation from which patients recover most rapidly.

Total hysterectomy
This is a much more common operation in which both the womb and the cervix are removed. It is the most likely operation for fibroids, endometriosis, prolapse or menstrual problems. In general the cervix will be removed automatically unless the woman for some reason does not want it to be and manages to persuade the surgeon not to do so – provided of course it is healthy.

Total hysterectomy with bilateral salpingo-oophorectomy
The womb, the cervix, both fallopian tubes and the ovaries are removed in this operation. It is usually performed to treat cancer of the uterus or ovaries or severe endometriosis or pelvic inflammatory disease. A total hysterectomy performed after the menopause may also involve removal of the ovaries so that there cannot be any further problems with them in the

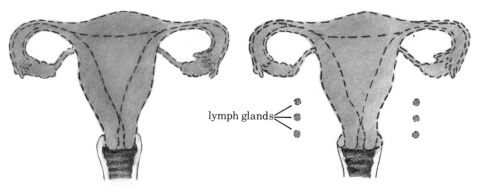

lymph glands

**Total hysterectomy with
bilateral salpingo-oophorectomy**

Extended or Wertheim's hysterectomy

future. Opinion is divided as to whether this is such a good idea since there is evidence that the ovaries continue to produce some hormones even after the menopause. Therefore, they may not be completely redundant after the menopause and their removal should be discussed by a woman and her doctor before this operation is carried out. Nevertheless, the ovaries *should* be removed if they are already diseased. If this operation is carried out on a pre-menopausal woman, she will be prescribed hormones to make up for lost production by the ovaries.

Extended or Wertheim's hysterectomy
In an extended hysterectomy, the top of the vagina is removed as well as the womb, fallopian tubes, ovaries and cervix. A Wertheim hysterectomy involves removing even more of the vagina, together with lymph glands and fatty tissue in the pelvis. This extensive surgery may be necessary if cancer has spread.

Vaginal or abdominal hysterectomy
Most hysterectomy operations are carried out by making an incision in the abdomen – either a horizontal, or bikini, cut just above the pubic hair, or a vertical incision down the centre of the abdomen. However, a few surgeons are expert in vaginal hysterectomy, whereby the uterus is removed via the vagina and this operation is becoming increasingly popular for women who have had children. The advantage of this operation is that there are no scars and, if all goes well, recovery should be more rapid; but it is a more difficult operation to perform and there is a greater chance of complications. However, vaginal hysterectomy may be a possibility if the uterus has already prolapsed, or dropped.

Going into hospital

Hysterectomy patients spend at least seven days in hospital but the usual length of stay is about ten days. As well as all the usual items such as nightclothes, toilet articles and books, it is a good idea to take some sanitary pads as most women have a slight discharge for a few days after the operation.

Going into hospital for any reason can be a frightening and confusing experience, and most people immediately feel very vulnerable and completely at the mercy of those looking after them. In a good hospital, doctors and nurses will be aware of the extra worries of hysterectomy patients and will be prepared to give all the necessary support. In other hospitals the woman may have to do a little more for herself to make sure that she gets all the care and emotional support that she needs. Making friends with other patients on the ward who have had or are about to have the same operation is a good way of building up confidence, although it is a mistake to be frightened or misled by things that other patients say about their treatment – their information may be inaccurate and alarmist. Those women who want more accurate information about the operation and other tests and procedures that may have to be carried out during their hospital stay, shouldn't be afraid of asking questions. It may be difficult to get answers without feeling aggressive or too demanding but it is worth persevering. Finding just one person, whether it is a junior doctor, nurse, physiotherapist or anaesthetist, who is easy to communicate with, can make all the difference. Women who have had any sort of problems with their operation or convalescence often say that they wish that they had asked more questions at the time.

The operation

A straightforward hysterectomy operation will probably not take more than an hour to perform. Coming round from the anaesthetic can be alarming but a nurse should be on hand. It may also be possible to have a relative or friend around at this time which can be very reassuring. Many people find that their throat is very sore when they wake up. This is due to a breathing tube that has been inserted into the throat during the operation.

A drip may be attached to the arm through which fluid and nourishment can be provided until the body recovers from the anaesthetic and is able to start digesting food normally again. And it is not unusual to find that a catheter has been inserted into the urethra to drain fluid that may accumulate in the bladder. Women who have not had a catheter inserted will be encouraged to pass urine within a few hours of the operation. Pain at the site of the surgery may be a problem but this can be easily controlled with the use of drugs.

Many women also experience considerable pain as a result of accumulated wind – the intestines are sluggish after the operation and it is not unusual to be constipated for up to three days. Colicky pains can be extremely uncomfortable – in fact some women find these pains worse than those caused by the surgery itself. But this problem will soon resolve itself, possibly with the help of a laxative or an enema. Walking round the bed as soon as possible is also one of the best ways to get wind moving.

It is also important to start moving about in the early days after the operation to reduce the risk of post-operative thrombosis, where a blood clot forms in the leg and travels through to the lungs. To prevent this most patients will be encouraged to get out of bed on the first or second day after

Getting out of bed after a hysterectomy
So as not to put a strain on the pelvic area when getting out of bed, slowly bend your knees by sliding your feet towards your buttocks.

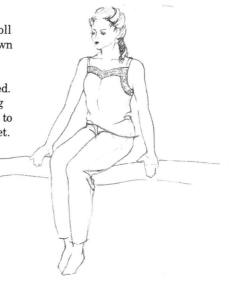

Keeping your knees bent, roll onto your side. Pushing down with hands or elbows, gradually swing your legs down over the edge of the bed. Once you are in this sitting position you should be able to reach the floor with your feet.

the operation. Getting out of bed can be difficult for the first few days and not all patients are given advice on the best way to do this when the pelvic area is still uncomfortable. First, the feet should be slowly pulled up towards the buttocks, so that the knees are bent. Then the woman rolls onto her side keeping the knees bent. The legs are then gradually swung down over the edge of the bed as she manoeuvres herself into a sitting position by pushing against the bed with hands or elbows.

Post-operative 'blues'

Many women feel very miserable two or three days after the operation, and find themselves crying for no particular reason. There are many reasons for this. It may partly be due to delayed shock from the operation; any sort of surgery can leave a person feeling down for a few days but gynaecological surgery, and particularly hysterectomy, is most likely to produce this kind of reaction. The stress of an operation can affect hormone levels making a woman more vulnerable to fear and depression, and she may feel a bit lost when the close attention given to her in the first few hours after the operation is withdrawn and she is expected to look after herself again. Sympathy and support from nurses and visitors should help most women through this difficult, but normal, part of the recovery process.

Exercise

A nurse or physiotherapist should help with exercises to tone up stomach and pelvic floor muscles after the operation. It is important not to do any exercises that cause pain but most women should be able to start doing pelvic floor exercises by the time they get home. Some women find the pelvic floor muscles difficult to identify but they are the ones used to stop urination in mid stream or to stop a bowel movement. All women should try to keep these muscles in tone by relaxing and contracting them at least twenty-five times a day.

Tummy-tightening exercises can be done while lying flat on the floor, with the knees bent, pulling in and holding tummy muscles for a count of four while pushing the small of the back downwards onto the floor. With practice this can also be done while standing up.

Sit-ups should not be attempted for at least ten weeks after the operation and these should be done very gently at first. Gentle stretching exercises started about six weeks after the operation are also important to prevent shrinking of the scar tissue. Begin by lying flat on the floor for ten minutes each day and progress, ten weeks after the operation, to lying on the stomach and slowly arching the back for six seconds. This should be repeated ten times, but discontinued if there is any pain.

Simple tummy-tightening exercises

Lie on your back with your knees bent and your arms by your sides. Pull in and hold your tummy muscles. Count to four and relax. Repeat five times. This exercise can also be done while standing.

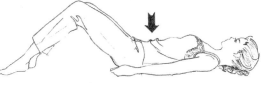

Using a cushion to support your head, lie on your back with your knees bent. With your hands resting on your thighs, lift your head up to look towards your knees and tuck your chin into your chest. Hold for a count of four then slowly relax, lowering your head. Repeat five times.

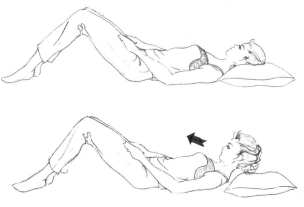

Coming home after the operation

If it is thought that a woman will not get adequate rest and care at home, she may be given the chance to spend the first two weeks after leaving hospital in a nursing home. The availability of beds in nursing homes varies according to the area, and also depends on whether consultants in a particular hospital think that convalescence in a nursing home is a good idea. Medical social workers in the hospital or health visitors should be able to give information on nursing homes if the consultant does not suggest it. Those who do go straight home, however, should not make any attempt to get back into their normal routine – at least two weeks' convalescence, with other people taking responsibility for the home and family, is usually necessary.

Most women feel exhausted for the first two weeks, but they should spend this time gradually building up their strength, for example by walking for a few more minutes each day. Nothing heavy should be lifted, so that means no strenuous housework for the first few weeks. Anyone living on their own with no one to help them should arrange for a home help for at least the first two weeks. After this period, a little more strenuous activity should be possible as long as there is no discomfort. Four weeks after coming home it should be possible for a woman to do gentle exercise.

The final checkup with the gynaecologist between six and eight weeks after the operation gives women a chance to ask any questions about the operation, and their progress since then. A woman should mention all worries at this time – anything left unresolved may turn into a more serious problem in the future. This includes physical problems such as pain or bleeding, and emotional problems such as depression or fears about resuming sexual intercourse. If all is well, the doctor will give the all clear which should boost confidence in getting back to a normal life.

Long-term recovery

The rate of recovery will depend on why the operation was carried out in the first place, what type of surgery was performed, fitness before the operation, and a woman's feelings about the operation. However, in general the more extensive the operation, the slower the recovery. For example, an extended or Wertheim operation for cancer may mean at least four months' recovery time, whereas a woman may feel fine two or three months after a total hysterectomy to treat fibroids.

However, every woman is different, and it is very important not to get dispirited by slow progress or to take any notice of other people's opinions on how quick recovery should be. Listening to your body is one of the best ways of deciding how much to do. It is important to do a little more each day, without overstretching the body. If exhaustion is a problem, sleeping or resting in the middle of the day is much more sensible than trying to catch up on the housework while the children are at school. If there is no one to help, then you should remember that a healthy, happy mother is much more important to a family than a tidy home.

Driving a car should be possible after the six week checkup, but to begin with care should be taken getting in and out of the car, and only short trips should be attempted for the first few weeks.

When to start work again will depend on the type of work involved. Highly stressful work or anything that involves a lot of physical exertion, standing, or travelling will not be a good idea for several weeks or months after the operation. A few women feel ready for anything a month after surgery but it is impossible to predict how long recovery will be so a woman should avoid committing herself to return to work on a particular date, if possible.

On the other hand it is worth remembering that activities such as driving and working should not be put off for too long as they are important in restoring confidence and independence which will all help a woman get back to her normal self again. At a certain point in the recovery process, sitting at home with not much to occupy the mind can, in many ways, be just as demanding as doing too much in the early weeks after the operation.

Possible problems during recovery

Any physical problems, such as bleeding after the six week checkup or pain, should be reported to the GP.

Some women find that several months after a hysterectomy they still have minor problems such as headaches, digestive problems or excessive tiredness which rarely affected them before the operation. The GP may be able to offer some help but most problems sort themselves out within a year or two, particularly if the woman takes lots of exercise and sticks to a good diet. However, those women who still find they have some irritating symptoms after this time, which have been adequately investigated by their doctor, should try to be positive about them and remember why the operation was done in the first place and what they have gained from it.

Hormone treatment

Women who have not reached the menopause and have their ovaries removed in the hysterectomy operation may experience some menopausal symptoms, such as hot flushes, mood changes and dry vaginal tissues, during their convalescence. Hormone replacement therapy will be given to make up for those normally produced by the ovaries but it still may take some time to get dosages right and for the body to adjust to this new situation. Some women who have not had their ovaries removed also report menopausal symptoms after a hysterectomy, but careful observation of symptoms and help from the GP should sort out these problems.

Depression

However well a woman has prepared herself emotionally for the operation she may still feel depressed, empty, useless or even angry. The best way to deal with this is to keep as busy as possible. Even if a woman has never done anything like it before, this may be the moment to get involved in local activities, evening classes or sports clubs. It may seem that none of this can help, but it is amazing how just small efforts in this direction can lead to great things.

It is also important to continue getting support from family and friends, by spending as much time as possible with those who are sympathetic, whether it be partner, mother or oldest friends. A combination of talking about problems and being distracted from them can be enormously helpful. If there is no one on hand to give this kind of support then contacting a local self-help group may be the answer.

Sex and sexuality

For many women, once they have got over the physical effects of the operation, the most difficult problem is getting back to a normal sex life, particularly if the hysterectomy had to be carried out before the couple were able to have children. In this case there will probably have to be a total reassessment of the marriage and the couple's hopes for the future, and professional help with this readjustment may be necessary from a marriage guidance counsellor.

However, many women find that the effect that hysterectomy has on their self-esteem, quite apart from the exhaustion it can produce, makes returning to normal sexual relations fraught with fears and problems, and may even be simply unappealing. If all is well most women will be told that they can start having intercourse again after the six week checkup. But if the woman does not feel like it, it is important for her and her partner not to try to pick up the physical side of their relationship just where they left off before the operation. They may have to take it very slowly to begin with, building up feelings of love and affection until both feel ready for full intercourse. The woman will need plenty of reassurance that she is still attractive and very much a woman even though she no longer has her womb. Any attempt to rush her when she does not feel emotionally ready or when she still feels some physical discomfort or pain will only delay her recovery. Women should remember that partners may also have fears about what has been done to the woman which they may have difficulty in talking about. In fact, the vaginal region will not be very much affected by the operation unless extensive Wertheim surgery has been performed in which the top of the vagina is removed.

Intercourse and orgasm may feel slightly different after the operation but it is not true that hysterectomy destroys sexual pleasure. The pleasure may be slightly different but with time and affection a couple should be able to adjust to the new situation. And in some cases the operation may improve certain aspects of sex by removing worries about pregnancy and contraception, taking away the inconvenience of periods and treating some problem that was making intercourse uncomfortable in the first place. A number of women find that intercourse and orgasm are much more pleasurable after hysterectomy.

If there are still problems after a few months the GP, marriage guidance counsellor or a hysterectomy support group should be able to help get things back to normal. And, in the few cases where a medical problem means that sex will never be quite the same again, it is worth remembering that intercourse is not the only way of giving and receiving sexual pleasure, and that there is much more to a good relationship or marriage than sex.

12 *Breasts*

by Jenny Bryan

Human breasts must be the most under-used mammary glands in the animal kingdom. In Britain sixty-five per cent of women start breast-feeding each of their 1.8 children but only one in five keep going for six months or more. Only in countries with very high birth rates do women's breasts come into their own. In Kenya, for example, where women have on average eight children, over ninety per cent of infants begin with mother's milk and nearly half are still being breast-fed at the age of twelve months.

Each breast consists of about twenty lobes where the milk is made, and each of these has its own duct which carries the milk to the nipple. When a woman becomes pregnant there is a big increase in levels of the sex hormones, oestrogen and progesterone, produced by the ovaries and placenta and the milk hormone, prolactin, which is released from the pituitary gland in the brain. These all help the breasts grow bigger and even small

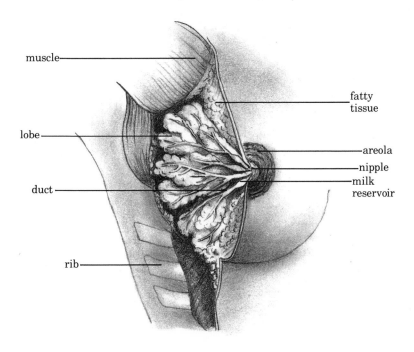

muscle

fatty tissue

lobe

areola

nipple

milk reservoir

duct

rib

When it is possible for a woman to breast-feed her child, it is beneficial to both mother and baby. However, breast-feeding has to be learned: the baby must be positioned correctly to suckle and the mother must feel comfortable and relaxed.

breasts can enlarge considerably during pregnancy – enough to breast-feed satisfactorily.

Milk production begins on a very small scale during the fourth month of pregnancy but really gets going only two or three days after delivery. Once feeding is under way, the baby's suckling triggers secretion of both prolactin and a second hormone, oxytocin, from the pituitary so that milk continues to be secreted into the milk ducts and out through the nipple. All this biological expertise for perhaps only a few months of activity in a lifetime!

The advantages of breast-feeding

Breast milk is the perfect infant formulation designed to give a baby all the nutrients it needs in the first weeks and months of life. In fact, the make-up of breast milk changes with the baby's age. Therefore, a premature baby will get more protein, calories and vitamins in its breast milk than a full-term infant. Moreover, its metabolic rate will be slower than a similar premature baby fed an artificial formula – thus enabling it to conserve more energy for growth.

Nutrients are not the only things transferred from mother to baby in breast milk. Breast-fed babies tend to have fewer infections than bottle-fed infants, probably because they receive their mother's antibodies in their milk. And there is some evidence that breast-fed babies are less prone to allergies. However, because so much is passed from mother to baby in this way, it is important for women who are breast-feeding to ensure that nothing harmful gets transferred. Ideally, it is best to avoid all medicines while breast-feeding – just as in pregnancy. Since this may not always be possible any woman planning to breast-feed should seek advice from her doctor about which drugs are safe to take and which are not. Clearly, if she must take medicines that are potentially harmful to her baby she will have to bottle-feed.

The advantages of breast-feeding are not all on the baby's side. The hormones needed for milk production tend to suppress those needed for egg production and ovulation. So women who breast-feed are less likely to get pregnant than those who do not. However, breast-feeding does not guarantee against pregnancy. Any woman who has just given birth, therefore, should discuss what method of contraception she will use once she starts having sexual intercourse again, although combined oral contraceptives should not be taken while breast-feeding since hormones may be passed to the baby through the milk.

Bottle-feeding

With the trend back towards breast-feeding – and the obvious advantages to the baby – many women feel they have failed their child if they cannot feed it themselves. This is most unfair. Breast-feeding is not as easy as falling off a log! It has to be learned and this takes time, care and patience. The baby has to be positioned correctly and the mother must feel comfortable and relaxed. Many women have problems initially and, particularly if they leave hospital within a day or two of giving birth, they may feel rather isolated. Health visitors are there to help and advise and can sort out many breast-feeding problems.

Even so, there are many women who cannot manage to breast-feed, either because they produce too little milk or because the baby simply will not suck properly. In these cases, rather than drive herself to distraction, it is much more sensible for a woman to transfer to the bottle. Modern infant formulas are very carefully balanced to include the nutrients that growing babies need. And no woman who finds she cannot feed her baby herself should feel a failure.

The breast as a sexual organ

If the only function of a woman's breasts were to feed her offspring, nature would indeed have been wasteful of its resources. Between their brief bursts of physical activity human breasts have a most important psychological role. They play an important part in a woman's image of herself. She may not feel the need for breasts like those of Raquel Welch or Samantha Fox but she probably expects them to conform broadly to the 'norm'. Not too big, not too small, not too high, not too low.

A woman's breasts will also play an important role in sexual arousal since they are highly sensitive to touch. Our errogenous areas may vary in detail but most women find stroking of their breasts pleasurable. Fortunately, they are not alone and men generally enjoy caressing their partner's breasts. The nipples and surrounding areolas are the most sensitive part of the breasts and the nipples slowly become erect during sexual arousal.

What is a normal-looking breast?

Women's breasts come in all shapes and sizes and, a bit like feet, one is often bigger than the other. As teenagers we may be jealous of the girls with the large full breasts but, as they begin to droop in middle age, it is often the smaller ladies with their upstanding breasts who have the last laugh. Small breasts are actually more common than you might think. One large chain store, for example, sells roughly equal numbers of its most popular bras – 34B and 36B.

Breast development begins for most girls at around the age of eleven or twelve and generally precedes the start of periods. There are variations and some girls develop quickly, others not until they are well into their teens. The actual growth occurs under the influence of the female sex hormone, oestrogen, levels of which start to rise before puberty. Breast tissue continues to be influenced by oestrogen levels – and not just during pregnancy. Many women who take the contraceptive pill, for example, which contains a synthetic form of oestrogen and progesterone, notice that their breasts get larger. This is generally reversible and the breasts return to their former size when a woman comes off the Pill.

Nipples vary not only in shape and size but in colour too. They can be anything from small, pale and flat through golden brown and pointed to large, round and reddish brown. A few women have inverted nipples which lie more like dimples on the surface of the breasts. There is nothing wrong with these though they may make it harder to breast-feed.

Breasts are made up entirely of fat and connective tissue; the only muscle lies much deeper on the chest wall. This means that as the skin becomes less elastic in middle and old age the breasts begin to sag. In some women this effect is accentuated or speeded up by breast-feeding – and during the 1950s and 60s this was one of the reasons why so many women preferred to bottle-feed. The breasts expand during pregnancy but, generally, they return to their normal size and, with the possible exception of a few stretch marks on the sides, look much the same after a baby is weaned as before the woman got pregnant.

Sagging can often be prevented by wearing a good support bra throughout pregnancy and breast-feeding if the breasts are particularly heavy. It is even worth wearing a bra at night. Sometimes, however, women lose so much tissue from their breasts after they stop feeding their baby that they do seem to be left with a couple of empty sacks. There is not a lot that can be done to prevent this but cosmetic surgery may be able to do something to correct it.

Changing what nature gave us

Some women become so depressed at the size or shape of their breasts – whether or not this has resulted from breast-feeding – that they go in search of a cosmetic surgeon to do something about it. In general, such operations are done privately. However, if a woman's mental state is considered to be seriously affected by distress over her breasts it may be possible for her to get surgery on the NHS. Many doctors are loath to refer women in all but the most severe cases. Frequently, they feel that women are blaming the shape or size of their breasts for much more deep-rooted problems – an unhappy marriage, general dissatisfaction with life, fear of growing old. Perhaps rightly, some doctors see breast surgery as no long-term answer to such problems.

Put simply, breasts can be either enlarged or reduced. A woman who feels her breasts are too small can have implants, generally made of silicone gel, through incisions underneath or at the sides of her breasts. Such operations have come a long way since the first football-like implants which were hard to the touch and bore little resemblance to the real thing. Today's implants are much more natural so that the breast is firm but soft and yielding. However, problems do still arise. Some women develop fibrous scar tissue around the implant so that the breast feels hard and immovable. This can be rectified by removing this fibrous 'capsule' around the implant, but success is not guaranteed.

Implants are also used to bolster up sagging, empty breasts after breast-feeding. Some of the flabby skin can be removed and the whole breast pulled back up into position when the implant is put in to fill it out. Alternatively, breasts may sag when they are very large and heavy. Reduction operations are performed quite routinely to remove some of the extra tissue and leave smaller, upstanding breasts.

Cosmetic surgery for breast enlargement or reduction is a skilled business. Unfortunately, there are many so-called 'cowboy clinics' run by people who are not qualified to perform such delicate and potentially mutilating operations. Anyone who wants to have this type of operation should be sure to seek out a reputable surgeon.

Ideally, a woman should begin by going to her GP and asking for a referral to a surgeon who is a member of the British Association of Plastic Surgeons or the British Association of Aesthetic Plastic Surgeons. If her GP refuses she will have to try to find such a surgeon for herself. Some private clinics advertise but their surgeons may not be members of the reputable associations. This does not make them incompetent but it makes it more difficult to find out if they are sufficiently skilled and experienced. Even in the most skilled hands breast surgery is no picnic.

The importance of breast examination

From the moment that increased oestrogen levels start to trigger breast development in her early teens the appearance of a woman's breasts varies during the menstrual cycle. Most women notice that their breasts swell slightly in the days leading up to a period and some find that their breasts seem rather lumpy at this time too. Both these effects disappear gradually during the period so that everything is back to normal at the end of it. For some women, however, the symptoms are much more severe. Their breasts feel tender and sore as well as swollen and lumpy. Often this forms part of the pre-menstrual syndrome when symptoms may include weight gain, feeling bloated and unaccountably angry or depressed (see Chapter 4).

It is because most women's breasts feel a bit lumpy before a period that examination of them should be done about a week after the end of menstruation. Women are now advised to examine their breasts for lumps or any other changes every month. This may seem a waste of time but it is very simple to do and may detect any unusual lumps and bumps at the very earliest stage so that they can be dealt with effectively. It is important to remember that any lump felt by a woman during self-examination is likely to be more easily treatable than any a doctor might find at a later stage. Above all, a woman should not worry about wasting the doctor's time if she finds anything unusual.

A simple guide to self-examination
Stand in front of a good size mirror and look carefully at the size and shape of your breasts and the position of the nipples. Stretch your arms up above your head and note any differences in the size or shape of your breasts. Then put your hands on your hips and push inwards so that you can feel your chest muscles tighten. Again, check to see if there are any differences

Breast self-examination

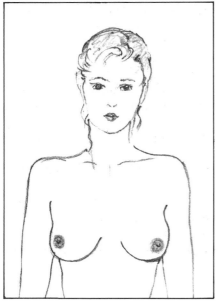 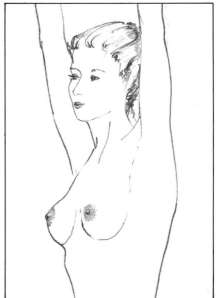

Looking:
This is to establish how your breasts normally look so that you will notice anything unusual. Stand in front of a mirror and look carefully at your breasts and the position of the nipples. Raise your arms above your head and note any changes in the size and shape of your breasts, turning slowly to look at the sides as well.

Feeling:
Lying on your back with your arms at your sides, examine each breast in turn by gently pressing in a circular movement (see inset). With your arm above your head, check the outside and underneath of your breast and the armpit for any signs of lumps.

in appearance. It is not unusual for breasts to look slightly different. The purpose of this first inspection is to establish how your breasts normally look so that you will be aware of any changes – bulges, lopsidedness, rash, discoloration – during later examinations.

Next, lie on your back with your arms at your sides. Decide which breast you will examine first and use the hand on the opposite side to circle it. Start at the top just above the nipple and press the tissue inwards to see if you can feel any unusual lumps or bumps. Move slowly round the breast in decreasing circles until you are encircling the nipple. Then do the same with your other hand on your other breast.

To examine the outside and underneath of your breast raise the arm on that side above your head before pressing the tissue with your other hand. With your arm still above your head move your hand up the side of your breast towards the armpit and feel for any signs of lumps.

Examining your breasts should not hurt – you should tell your doctor if any part is continually tender or painful. Equally, gliding your hand over the surface without pressing the tissue will not show anything up. So be firm but gentle!

What if I find a lump?

Nine out of ten lumps that women discover in their breasts during self-examination are benign. That is the most important thing to remember if you think you have found something abnormal. It does not mean that you can ignore a lump; go to your GP as soon as you can so that he or she can examine your breasts and decide whether you should see a specialist.

The most common causes of lumps in the breasts are cysts and fibroadenomas. Cysts can form when one of the milk sacs or ducts becomes swollen and filled with fluid. To the practised hand of a breast specialist they will feel soft and pudgy as the fluid moves when the tissue is pressed. Fibroadenomas are collections of fibrous, glandular tissue and they will feel harder and more solid to the touch.

What happens next?

If the GP agrees that there seems to be some sort of lump you will be referred to the outpatients' clinic of your local hospital to see a specialist. He or she will also examine your breasts and feel under your arms and at the base of your neck to see if your lymph nodes are enlarged. Lymph nodes are the 'stations' within the lymphatic system of ducts and tubes that carry fluid to the tissues around the body. They are responsible for getting rid of excess fluid from the tissues or topping up levels when we are dehydrated. The lymphatic system also plays a part in producing immune cells and the lymph nodes are often swollen when we are feverish and have an infection. Cancer cells use the lymphatic system to spread to other parts of the body which is why the breast specialist will be keen to see whether they are swollen. This does not automatically mean that the cancer has spread – just that the lymph nodes, as well as the breasts, will need treatment.

The specialist will want to take a sample of cells from the breast lump so that they can be examined in the laboratory under a microscope. How the sample is taken will depend on how the lump feels. If it is soft and spongy and clearly filled with fluid, a fine needle will be inserted into the lump and fluid drawn off through a syringe. This is called needle aspiration.

If the lump feels harder – like a fibroadenoma – a sample will be taken under local anaesthetic. Once again, a needle is inserted to remove the cells and this is called needle biopsy. Sometimes, the specialist prefers to carry out a more extensive operation to remove the entire lump under general anaesthetic and this will mean an overnight stay in hospital. This procedure is called an excision biopsy.

As well as taking samples from the breast it is usual to perform X-rays to get a better picture of the lump. This is called mammography. It differs from a chest X-ray in that the rays pass downwards through the breast rather than being directed from the front through the chest. Mammography is now so accurate that it can pick up the very first signs of a tumour even before the lump is properly formed and can be felt with the hands, since it shows up as groups of calcium deposits in one of the milk ducts. However, regular screening in this way is currently not carried out on women under the age of thirty-five.

Breast X-rays are often used to help the surgeon plan how to remove the lump. If, for example, the tumour is unlikely to be visible to the naked eye during surgery wires may be passed into the breast, guided by the picture on the X-rays, so that they converge in the area of the calcium deposits. During the operation the surgeon finds the wires and removes all the tissue in the surrounding area. A second X-ray can then confirm whether or not all of the abnormal tissue has been removed.

When the tests show that it is cancer

If the cells sent to the laboratory turn out to be cancerous the next step is to remove the entire lump from the breast. If an excision biopsy has been done, the lump will already have been removed and, depending on the extent of the tumour, the doctor will advise whether further surgery is needed or some other form of treatment.

It is possible for the lump from an excision biopsy to be examined immediately and, depending on the results, further surgery done straight-away while the woman is still under general anaesthetic. Before going into the theatre you should be absolutely clear what your surgeon plans to do – whether or not he will act immediately on the results of the excision biopsy, possibly removing the whole breast if the lump is cancerous.

Some women prefer to leave it to the surgeon to decide what to do while they are under anaesthetic and prepare themselves for the fact that they may come back from theatre without their breast. If, however, you want to discuss the results of the laboratory tests before deciding on further surgery make sure that the surgeon knows that you are consenting *only* to the excision biopsy and not to further surgery during the initial operation. It is sad but true that women are still being wheeled into the operating theatre unaware that they might have cancer and wake up back on the ward minus one of their breasts.

In recent years, operations for breast cancer have become less extensive and disfiguring than even five or ten years ago. This is because doctors are discovering that in the vast majority of cases a woman's chances of survival are as good with a small operation to remove just the cancerous lump as they are when the whole breast is removed (mastectomy).

There are exceptions. Rarely nowadays does a woman delay going to the doctor for so long that the tumour spreads widely through her breast and it is necessary to remove the whole thing. Also, it may not be possible to leave any tissue if there is a large tumour in a very small breast so a mastectomy is the only option. There is a small risk that a tumour may recur if only the lump, and not the whole breast, is removed, though this risk is reduced by radiotherapy after the lumpectomy. However, if there is a recurrence mastectomy is likely to be the only answer.

Some women prefer to have a mastectomy at the outset, rather than a lumpectomy, because they feel that is the safest thing to do. By getting rid of all the tissue that may have been in contact with the tumour they feel they can put the whole episode behind them. Whether or not a woman chooses to have a mastectomy, the most important thing is that she is the one who has made the decision. Obviously, she will want to listen to and take the advice of her doctors, but the final decision should be hers.

Anyone who is unhappy with the advice they are given by their doctor can ask for a second opinion. Unfortunately, this is not an automatic right. However, any reasonable doctor, confident of his or her skills, should not feel threatened by a patient asking for a second opinion and should be quite happy to refer to someone else in the same or a different hospital. Anyone who has difficulty getting their specialist to refer them for a second opinion can go back to their GP and ask for a referral note.

If you decide to seek a second opinion you should be aware that this may add to your dilemma especially if the doctors give conflicting advice. It is possible even to get a third opinion. But time will be ticking by and, although with breast cancer a delay of a few days or weeks between diagnosis and treatment is not dangerous, it is obviously important to make a decision and have the cancer treated as soon as possible.

After a lumpectomy

It is usual to have a course of radiotherapy after lumpectomy to reduce the risk of the tumour coming back. Radiotherapy treats cancer with high-energy rays that destroy tumour cells. It can be given in one of two ways. A course of external radiotherapy can be given once or twice a week over several weeks. This means lying under a large piece of equipment while the rays are directed very specifically at the breast from which the lump was removed. Alternatively, it is possible to insert radioactive wires into the breast at the same time that the lump is removed. These slowly release radioactive material over the next few days to destroy any remaining cancer cells. After the treatment is finished the wires are removed and the woman goes home. Radiotherapy has fewer side-effects than most forms of cancer treatment but people often feel tired and sick and lose their appetite. However, all of these problems disappear once treatment is stopped.

During the operation to remove a breast lump the surgeon will also examine the lymph nodes under the arm and in the neck for any signs of cancer and probably remove some lymph tissue for examination in the laboratory. If tests show that the cancer has spread to the lymph nodes a woman is generally advised to have a course of chemotherapy.

A number of anti-cancer drugs are available and because they are injected into the blood-stream they can travel round the body and destroy cancer cells that have not been removed either by surgery or radiotherapy. Courses of treatment can last several months and unfortunately side-effects can be rather unpleasant – including sickness, hair loss and feeling generally unwell. All these side-effects stop after treatment and hair grows back; in addition, there are simple steps to minimize side-effects which should be explained at the hospital.

Recently, doctors have found that women whose tumours are found to be sensitive to the female hormone, oestrogen, can benefit from additional anti-hormone treatment. The most commonly used drug is called tamoxifen. This treatment is slowly replacing the need for women to have their ovaries removed if they have a recurrence of their breast tumour since it is the ovaries that produce oestrogen in women of reproductive age.

After a mastectomy

A mastectomy is a major operation and, depending on her age and general health, a woman is likely to spend about ten days to two weeks in hospital after surgery. Initially, a tube will drain fluid from the place where the breast has been removed and most women need some painkilling drugs to relieve any discomfort. If the lymph nodes in the armpit have been removed as well as the breast, the arm on that side is likely to feel a little stiff and a physiotherapist will probably be brought in to start some exercises. The drainage tubes come out after a few days and the stitches about a week later. By that time most women are well enough to go home to continue their convalescence. Further treatment will depend very much on the extent of the cancer. Doctors may well advise courses of radiotherapy and/or chemotherapy and anti-hormone treatment similar to those following lumpectomy.

While the scar is still healing a woman is given a soft, lightweight prosthesis to slip into her bra so that her breasts will look the same size when she leaves hospital and she will not feel self-conscious. This prosthesis, called a Cumfie and originally designed by the Mastectomy Association, is only temporary.

Once the scar has healed a woman can choose a more permanent prosthesis from what has become an enormous range of breasts in every conceivable size and shape. Some mimic the shape of the breast itself. Others have a small extension for women who have had more tissue removed or a larger extension if tissue has been taken away between the breast and the armpit. There are also mini-prostheses to build up the breast for women who have had only a small part removed. Prostheses are made from a variety of materials but those filled with silicone are the most popular as they closely resemble the shape and feel of a natural breast.

Any woman who has had a mastectomy is entitled to a new breast prosthesis each year on the NHS. And if she loses or gains a lot of weight she can also have a new one to match her new size. It is very important that a woman should feel confident with her prosthesis and for this reason it is worth insisting on seeing and experimenting with a wide range. It is quite unacceptable for any woman simply to be given a box of prostheses and left to get on with it. Many hospitals have qualified prostheses fitters and they should have a good selection to choose from.

Some women assume that they must cover up after a mastectomy and that they must give away their pretty clothes – sundresses, swimming costumes and low-necked blouses. This simply is not true. There is no reason why anyone other than the woman and her partner – and those she chooses to tell – should know that she has had a mastectomy. Some manufacturers design swim and beachwear specially for women with mastectomies and a few extra stitches or safety pins mean that even the most self-conscious women can wear low-cut clothes. The Mastectomy Association produces a number of leaflets for women who have had a breast removed and these include advice on adapting clothes to save embarrassment.

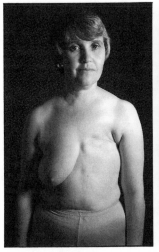

After a mastectomy operation women are helped to choose and fit a prosthesis. This can be slipped into her bra, and has the shape and feel of a natural breast. Once the woman is dressed, it is impossible to tell that she has had a breast removed.

Breast reconstruction

A growing number of women are having operations to reconstruct their breast after surgery for cancer. And there is a move towards all women being given the option of reconstructive surgery either at the time of mastectomy or later on.

There are a variety of operations but increasingly popular is the technique of stretching the remaining skin on the woman's chest to accommodate a silicone implant. A temporary bag is put in under the remaining chest tissue and slowly pumped up with injections of fluid. This is done over a period of several weeks so that the skin is stretched to accommodate an implant. When this is done the bag is removed and replaced with the silicone implant. Often it is possible to reconstruct a nipple from tissue elsewhere in the body, though some women do not bother.

The advantage of this type of operation over other reconstructive surgery is that it uses the woman's chest skin to cover the implant so that the texture and colour of the new breast matches the other. If skin is grafted from other parts of the body – such as the leg or abdomen – the match will not be as good since skin varies in appearance depending on where it has come from. Also, there will be scars where the skin has been removed for the purpose of grafts.

This is not to suggest that the 'stretch' operation is an easy option. The process can be quite painful and any woman embarking on this sort of reconstructive surgery should be fully aware of what she is letting herself in for – the chances of success and the risks of failure. A reconstructed breast will never be quite the same as a natural one. It will have no sensation because there will be no nerve endings in the nipple and, of course, it will not produce milk or be used for feeding a child.

Accepting the loss of a breast

The vast majority of women walking around today after a mastectomy have not had a breast reconstruction. Either they were unaware of the possibilities of surgery or they decided against it. How well women accept the loss of a breast varies enormously. Some take it in their stride – glad at least that the cancer has been taken away and they can get back to normal life. Others find it much harder to accept. Clearly, the loss of something as critical to her body image as a breast is traumatic for any woman.

It is not surprising that some women feel that they have become ugly and undesirable – both in their own eyes and those of their family and friends. How quickly they come to terms with their loss will depend to a large extent on the love and support they get from those close to them. While still in hospital women are encouraged to look at their scar after a mastectomy. This should not be forced on them and it should be in their own time. But studies have shown that it is women who have the most difficulty in accepting their scar who have trouble recovering emotionally from their operation. It is important not to build up the first look into a major event but for a woman to take it in her stride as part of the process of recovery.

A woman is naturally going to worry about the reactions of her husband or partner. However, it is the experience of the Mastectomy Association that these worries are groundless and that women worry far more than their partners about the loss of the breast. A strong relationship is not going to break down because a woman has a mastectomy and, if there are problems as a result of the operation, these would probably have occurred sooner or later anyway.

It is most likely to be women who do not have a regular partner at the time of their mastectomy who are faced with the biggest dilemma. How do they cope when they do meet someone new? Obviously it is not something you discuss at the start of a relationship but is there ever a right time? And how is the new partner going to cope with the news that his girlfriend only has one breast. As you might expect, women's experiences vary. Some partners accept quite easily that the woman they are becoming close to has only one breast. They were attracted by more than just her breasts! And the relationship is with her whole person – personality, sense of humour, arms, legs, face, the lot! Other men do drift away. Perhaps their attraction was not strong enough and the relationship would have been a physical rather than an emotional, loving and caring one. In that case, it is probably better to find out early on than to become too attached. There are no easy answers and inevitably there is a lot of heartache.

The medical profession recognizes that mastectomy carries as many if not more emotional than physical problems. Some breast clinics include nurse counsellors as part of the team caring for women with breast cancer. It is part of their job to support women undergoing treatment – to help them talk through their fears and to give them practical advice about the services that are available to help. Many women benefit from talking to other women who have already had a mastectomy and dealt with all the problems facing the newer patient. Some hospitals arrange such meetings and the Mastectomy Association can also put women in touch with each other. It is not for everyone – but it can help enormously.

Breast cancer: those most at risk

Cancer specialists are beginning to find that some groups of women are especially prone to breast cancer. Very few types of cancer run in families but breast cancer does appear to be an exception. There is no direct genetic link but women with a mother or sister who has had breast cancer do seem to be at greater risk of getting the disease too.

Also at greater risk are women who do not have children, or women who start their family later in life. Certain patterns of breast tissue seem to be more common in women who get breast cancer. And women with low levels of a chemical called sex hormone binding globulin (SHBG) may also be at higher risk. This is because this chemical is responsible for transporting and carrying around the female hormone oestradiol. Low levels of SHBG means that larger amounts of oestradiol are travelling around uncontrolled and these seem in some way to be associated with breast cancer.

There has been enormous debate about whether the contraceptive pill can contribute to breast cancer, particularly if it is taken by young women whose breast tissue is still developing. The results of the major studies that have taken place so far have been conflicting and it could be several more years before there is any definitive answer.

In the meantime it may be wisest for women in their mid to late teens to avoid taking the Pill if at all possible. And women who have taken the Pill for periods of longer than eight years may also prefer to discuss alternative methods of contraception with their doctor. No drug is free from side-effects and in deciding whether to take the Pill it is important to weigh the benefits against the risks. The link with breast cancer is, as yet, unproven. Yet there can be little doubt about the emotional turmoil for a young single woman who gets pregnant and has to decide whether to have an abortion, which carries its own small risk.

Other methods of contraception also carry their own hazards; in general, for example, women who have not had children are sometimes advised against an intrauterine device although this is not a hard and fast rule. So any woman worrying about the theoretical risk of breast cancer from the contraceptive pill must also consider the risks of alternative methods of contraception (see Chapter 5).

Can we avoid breast disease?

Until more is known about the way in which benign and malignant breast diseases develop we probably cannot do very much to avoid them, except be alert to the earliest warnings of something wrong and seek medical help.

The government continues to consider how to set up an effective breast screening programme and in the meantime facilities for routine mammography outside the private health system are few and far between. At present a woman can only get a breast X-ray on the NHS if she is thought to have some kind of lump or if she has a strong family history of breast cancer. Several large studies are under way to find out whether regular routine breast X-rays, every few years, really can pick up more breast tumours at an earlier stage and enable them to be treated more effectively and improve survival rates from breast cancer. In the meantime, regular self-examination of her breasts gives a woman the best possible chance of early diagnosis and treatment.

There is no doubt that progress is being made in the treatment of breast cancer – not least in reducing the need for mutilating surgery. The goal must now be to improve long-term survival rates for women with the disease so that breast tumours can become one of the success stories of cancer therapy.

13 *Special Problems*

by Antonia Rowlandson

Painful intercourse

Pain during intercourse (or dyspareunia) may be felt at the entrance to the vagina or deep in the vagina and abdomen. There are many possible causes of discomfort during intercourse, but a woman should always see her doctor if this pain persists. Minor problems such as infection are often easily treated and more serious problems, which may be uncovered during investigation of pain during intercourse, should be attended to as soon as possible after discovery.

Pain at the vaginal entrance may be due to infection, inflammation, or some other problem in the external genital area or vagina. Vaginal dryness is also a common problem, particularly if a woman is not sexually aroused, or if she is going through the menopause. Lubricating jelly or hormone cream may help in these circumstances.

Deep pain during intercourse may be due to problems such as pelvic infection, fibroids, endometriosis or abnormal positions of the womb. Discomfort may also be experienced after gynaecological operations and childbirth, particularly if there has been tearing of tissues during childbirth or if an episiotomy, or small cut, has been made in the tissues to prevent tearing. But the doctor should be able to help if the damaged tissues do not heal satisfactorily and continue to cause pain during intercourse.

Painful intercourse may also be due to psychological problems. If a woman is frightened of sexual intercourse, or believes that sex is wrong or fears that she may get pregnant, or cannot become aroused, intercourse may be difficult and painful. The muscles of the vagina may even go into spasm – a condition called vaginismus – making penetration impossible. Sympathetic help from a doctor, psychotherapist or sex therapist should resolve these problems, allowing a couple to enjoy a normal sex life.

Fibroids

Fibroids are benign tumours made up of fibrous and muscular tissue that usually develop within the muscular walls of the uterus. They are very common – it has been estimated that one in five women over the age of thirty-five has fibroids, although in most cases they are small and do not cause any problems. It is not known exactly why fibroids develop but because they are usually seen during the reproductive years of a woman's life, especially in women who are taking the Pill, and because they tend to become larger during pregnancy and decline after the menopause, it has been suggested that they are associated with oestrogen activity.

Symptoms

The symptoms are variable – it is not uncommon for a large fibroid to cause no problems at all, whereas small ones may give rise to distressing symptoms. Heavy periods are the most common symptom and if not treated this can lead to anaemia. Pain, abdominal swelling and urinary problems are other possible symptoms, and occasionally recurrent miscarriage may be due to the presence of fibroids.

Diagnosis and treatment

Fibroids may be diagnosed by X-ray, ultrasound or by an exploratory operation. Treatment of fibroids is not necessary unless they are causing problems. Drug treatment with hormones is sometimes suggested but, if heavy bleeding is the main problem and the fibroids are small, a D and C may be the solution.

However, an operation may be necessary for larger fibroids and more severe symptoms. Surgery involves either removing just the fibroids themselves, or performing a hysterectomy in which the whole uterus is removed. From a surgical point of view, hysterectomy is a more straightforward operation and also eliminates the possibility of fibroids returning in the future; but it is more traumatic for the woman so local removal of fibroids should be carried out if possible, particularly if a woman plans to have children.

Endometriosis

This is a condition in which tissues resembling the endometrium, or lining, of the womb grow either within the muscular walls of the uterus or in other sites outside the uterine cavity such as the ovaries, lower genital tract or bowel. The causes of endometriosis are not completely understood but it is possible that in some cases it is due to a backflow of menstrual blood, containing endometrial cells, up the fallopian tubes and into the pelvic cavity. These endometrial cells attach themselves to other sites and tend to bleed when normal menstruation occurs, becoming congested and painful. Depending on the site of the condition, there may also be backache or rectal pain, or pain during intercourse. Periods may also become irregular and endometriosis can be a cause of subfertility (see Chapter 9).

Diagnosis and treatment

It may be possible to diagnose some cases by examining the internal organs using a laparoscope, an illuminated tubular instrument that is passed through a small incision in the abdomen. But surgical exploration may sometimes be necessary.

The only real cure for women with severe symptoms is hysterectomy and removal of the ovaries. The woman will immediately experience menopausal symptoms and will be given hormone replacement therapy to replace the oestrogen normally produced by the ovaries.

Hormone treatment or less radical surgery may be the answer for young women who want to have more children. Symptoms of endometriosis disappear during pregnancy and some women find that the condition does not come back after the birth. More information can be obtained from the Endometriosis Society (see Useful Addresses).

Prolapse and incontinence

Prolapse of the uterus and vaginal walls is a fairly common gynaecological complaint, particularly in women who have had children. In this condition the tissues that support the uterus and vagina are not strong enough to hold them in place and the organs begin to sag downwards, possibly exerting pressure on the bladder and causing stress incontinence. This means that urine has a tendency to escape when there is a physical stress such as coughing, sneezing or jumping.

Some women are born with weak pelvic floor muscles, others may find that problems start after having children. Hormonal changes during the menopause may also make a woman more vulnerable to prolapse. But there is no doubt that the condition tends to get worse as a woman gets older, and that exercise is the best way of preventing prolapse and avoiding the associated urinary problems. Pelvic floor exercises which are often

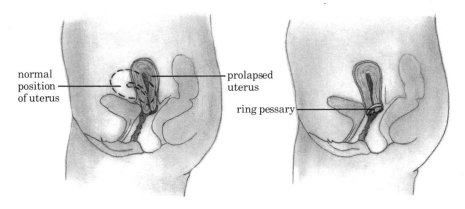

normal position of uterus — prolapsed uterus

ring pessary

taught in antenatal classes are very useful, and vital if the pelvic floor has begun to sag. But any exercise, including walking, jogging, swimming or yoga, will help to prevent the weakening of these important muscles.

The first symptoms of prolapse may be a feeling of discomfort, a sensation that something is coming down in the vagina. There may also be urinary symptoms such as a need to pass urine frequently or leakage of urine. Another sign of prolapse is that feelings of discomfort tend to be worse when a woman stands and get better when she lies down.

Treatment may involve surgery. Straightforward repair of the pelvic floor muscles is usually very successful, but more severe cases may be treated with a vaginal hysterectomy, particularly if the woman has completed her family. Young women with mild symptoms or elderly women

Yoga positions
Inverted yoga positions are particularly helpful for reversing the effects of gravity and thus for treating prolapse.

Shoulder stand
Lie on the floor with your legs bent. Lift your knees towards you and use your hands, pushing down on the floor, to lift your hips. Bring your legs up over your head at an angle of about 45°. Straighten your legs and spine supporting your back with your hands. Breathe deeply and hold the position for as long as is comfortable. Slowly roll out of the position.

The plough
Lying on the floor, slowly lift your legs until they are stretched out straight above you. Lower your legs over your head behind you until they touch the floor, supporting your body with your hands. Hold for as long as is comfortable, then slowly uncurl.

who are not fit enough to have surgery may be fitted with a ring pessary – a flexible plastic ring inserted into the vagina to support the uterus.

Urinary incontinence
This is not only caused by prolapse. Other possible causes are infection, problems with the bladder or urethra, or psychological problems such as anxiety. But incontinence is a very common problem among women of all ages and much can be done to help. Some areas now have clinics specifically designed to help with incontinence problems and any woman who has trouble getting satisfactory treatment or who would like more information should contact the Incontinence Advisory Service (see Useful Addresses).

Ovarian cysts

Cysts on the ovaries can arise when cells of the follicle become enlarged and filled with fluid instead of releasing an egg. Cysts can grow quite large without causing any problems but possible symptoms include pressure on the bladder and rectum, pain during intercourse, or abdominal swelling and tenderness. In some cases they may also cause periods to become irregular or stop altogether. Cysts may be felt by a doctor during a routine pelvic examination and ultrasound can then be used to confirm the diagnosis.

Small, symptomless cysts will probably not require treatment as often they cause no problems and may even disappear. But larger cysts that do not disappear should be removed surgically to make sure that they don't cause problems in the future. Surgery will also be necessary if there are complications such as torsion or twisting of the cyst, or rupture or infection.

It may be possible to remove the cyst without damaging the ovary but in a few cases it may be necessary to remove the whole ovary as well. But this shouldn't cause any problems provided the other ovary is functioning normally. If the woman is approaching the menopause, the gynaecologist may suggest a hysterectomy involving the removal of both ovaries and the womb, to exclude the possibility of further cysts or of cancer of the ovaries or womb in the future.

Screening for cancer of the cervix

Cancer of the cervix kills approximately 2,000 women in England and Wales each year. The number of women under thirty-five dying from the disease has increased dramatically over the last few years but this group still only accounts for less than six per cent of all deaths from cervical cancer. It is now well known that regular smear (Pap) tests to check the condition of the cells of the cervix can detect precancerous abnormalities and that treatment in these early stages is quick, painless and successful.

The number of deaths could be dramatically reduced if all sexually active women had regular smear tests. But despite much publicity about the disease and screening for it, and many smear tests (more than 30 million have been carried out in the UK since the test was introduced twenty years ago), there has only been a small reduction in the incidence and the death rate from the disease in this country. This may be due to an increase in the incidence of the disease over the last few years but it is also probably because there are still many women at risk who have never had a smear. A large proportion of the deaths from cancer of the cervix occur in women aged between forty-five and sixty-four, but many of these have not been screened, partly because older women are less likely to attend family planning and antenatal clinics where screening is carried out routinely, and partly because they often dislike internal examinations and fear what the doctor may find.

How often should a woman have a smear test?

There is some disagreement about how often a woman should have a smear test. However, many experts agree that the first test should be done as soon as a woman becomes sexually active and, if it is negative, further tests should be done once a year for the following two years. After that, and provided the tests are still negative, a woman should have a smear every two years. Annual smears should be offered to women who have had an abnormal smear in the past and smears will also be taken from pregnant women and from those with gynaecological or venereal symptoms. Women over sixty-five do not need to go on having smears if the last two have been normal, but if there are any abnormalities they will be advised to go on having tests until there have been two successive normal ones.

The factors that are thought to increase the chances of developing the disease are:

Early sexual intercourse
Women who start to have intercourse at an early age may be at increased risk because the young cervix may be more vulnerable to agents that encourage the development of cancer.

Number of sexual partners
Women who have many sexual partners or whose partners have had many sexual relationships are more likely to come in contact with cancer-causing agents and so are at greater risk.

Infections
At the moment researchers are investigating the apparently strong associ-

ation between the wart virus and cervical cancer – there is often evidence of the wart virus in women who have cervical cancer or early signs of the disease. Anyone, male or female, who has warts in the genital area should see their doctor as they can often be easily cured. But venereal disease and herpes have also been suggested as having some role in the development of cervical cancer, and women who have had these infections are advised to have annual smear tests.

Smoking
A woman who smokes 20 cigarettes a day is thought to be seven times more likely to develop cervical cancer than a non-smoker.

Contraception
There is no direct link between taking the Pill and developing cervical cancer but women who do not use barrier methods of contraception, such as the cap or sheath, are more likely to be exposed to infections such as the wart virus, so women who take the Pill or have an IUD should perhaps have more frequent smear tests.

Where to go for a smear test

Free smear tests are now widely available – in fact, some experts are concerned that the enthusiasm with which they are carried out by some doctors and clinics has meant that many young women have been tested unnecessarily, often while many older women have never been screened.

GPs, family planning clinics, and centres that carry out well-woman checkups will all do smear tests. A smear test is also always carried out as part of a private health checkup. Some areas have set up computerized screening programmes that recall women for smears at regular intervals and arrange for results of tests to be forwarded. Those who are not involved in one of these schemes should keep a record of when they had their last smear and when the next one is due, and should ask for the result of the test if possible. Usually the woman will not be told if the smear is normal.

Having a smear

Great skill is required to collect a satisfactory sample of cells from the cervix and technicians who examine the cell samples for abnormalities report that many are not good enough for a proper assessment to be made. Consequently, another sample of cells has to be taken.

A woman should try to have her smear test carried out by a doctor or nurse who is skilled in the technique – a well-woman or family planning clinic would probably be the best place to go. Some women find that having

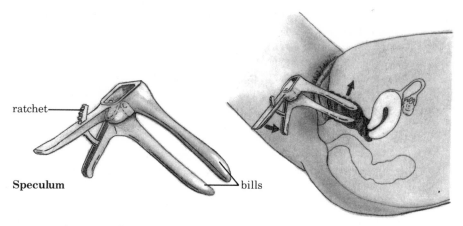

ratchet

Speculum   → bills

a smear is uncomfortable, but it shouldn't be painful. As in most gynae-cological examinations a metal instrument called a speculum is inserted into the vagina, allowing the doctor to have a clear view of the cervix and vaginal walls. He or she can then collect a sample of cells from the cervix by gently wiping a spatula across it.

Cancer is most likely to develop in an area called the squamo-columnar junction where soft, columnar cells in the cervical canal which leads to the uterus are gradually changing into relatively hard, squamous cells on the part of the cervix that is exposed to the vagina.

Results of the cervical smear

Very few women are found to have abnormal cells in the sample taken from the cervix. Moreover, an abnormal smear does not mean that a woman has cancer, although she may be advised to have further investi-gations and possibly some simple treatment to prevent the development of cancer in the future.

Many medical terms may crop up in the description of smear results, some of which can sound rather alarming. These are some of the descriptions that are most likely to appear together with the appropriate action that will be taken to deal with them:

A *normal* or *negative* smear means that no abnormality was found. The woman should have another test in 1-2 years' time depending on when she had her first smear and on whether any of the risk factors apply to her.

Cervical erosion: This means that some of the soft columnar cells usually found in the canal leading up to the uterus have become exposed to the vagina and appear red and inflamed. This is a fairly common condition, especially in women taking the Pill and symptoms include a discharge or bleeding after intercourse. If the cells do not return to normal in a few months or the symptoms become distressing, the troublesome area can easily be treated by freezing the cells.

No endocervical cells seen – negative: This tells the doctor that the sample did not contain enough cells from the area of the cervix where cancer is most likely to develop with the result that the person examining the cells was not able to exclude definitely the possibility of abnormal cells. A further smear test will have to be carried out.

Atypical means that the cells show some very slight abnormalities. This can be due to changes in the cervix that occur throughout the menstrual cycle or to an infection such as trichomonas. A smear test may also pick up signs of the wart virus which is thought to be involved in the development of cervical cancer. The doctor may recommend diagnosis and treatment of any infection and a repeat smear test in a few weeks' time.

Mild dysplasia (or *CIN 1,* which stands for cervical intraepithelial neoplasia stage 1 – changes in the cells of the cervix that precede the development of cancerous cells): This means that the cells show evidence of precancerous abnormality but treatment may not be necessary as the cells may return to normal. A repeat smear will probably be carried out within the next few months and then annually.

Moderate dysplasia (CIN 2) and *severe dysplasia/carcinoma in situ (CIN 3):* These more severe precancerous abnormalities should always be investigated further and treated as the cells are unlikely to return to normal on their own.

Abnormal smears should first be investigated using a technique called colposcopy. This is an outpatient procedure in which a binocular microscope is used to examine the cervix. It should not be painful and sympathetic treatment should minimize any discomfort. It may be necessary to carry out a biopsy during the colposcopic examination. This simply means that a needle is used to take a small sample of cells for further examination. Doctors will then use the results of these investigations to decide on the best treatment. It is now widely acknowledged that no patient with an abnormal smear that suggests the presence of precancerous cells should be treated without first being investigated with colposcopy and possibly biopsy, but unfortunately in some areas there is still a shortage of equipment and properly trained colposcopists. A woman may therefore have to ask to be transferred to a colposcopy unit in another area if there is not a local one.

Treatment of abnormal cells

If the doctor decides that the abnormalities are suitable for treatment, the abnormal cells will be destroyed by one of several methods. Cryosurgery can be used to destroy small areas of abnormality by freezing the unwanted tissues. Electrodiathermy, usually done under general anaesthetic, uses heat to destroy abnormal cells. But laser treatment, which can be given in an outpatient clinic, is now seen as one of the best ways of getting rid of abnormal cells. It is successful in ninety-seven per cent of cases and causes little discomfort though there may be some bleeding or discharge after treatment. However, healing should be quicker when this method is used.

In the 1970s cone biopsy, which involves removing a cone-shaped piece of tissue from the cervix, was the usual treatment for precancerous cells in the cervix. But this treatment can damage the cervix causing complications in pregnancy and childbirth. Today, therefore, doctors favour the more conservative treatments which give rise to fewer problems, particularly in the case of young women who hope to have a family in the future.

After treatment for an abnormal smear a woman should have two more smears at yearly intervals and, if these are normal and she has no other risk factors, she will probably be recommended to return to screening every two or even three years.

Invasive cancer of the cervix

The cervix is the most common site for cancer in the female genital tract but it is still a rare disease responsible for under four per cent of all cancer deaths in women. The disease can be cured in many cases particularly if it is diagnosed in the early stages – seventy-five per cent of cases in the first stage of cervical cancer survive at least five years.

The first sign of the disease may be postcoital bleeding or bleeding between periods, or, in the case of postmenopausal women, after periods have stopped. There are many possible reasons for unusual bleeding – for example, cervical erosion or the method of contraception used such as the mini-pill or the IUD – but full investigations of symptoms should always be made as cancer is a possibility.

If cancer is diagnosed, treatment will depend on the extent of the cancer as determined by investigations such as biopsy and ultrasound, and on the age of the patient and whether she is likely to want children in the future. Surgery, possibly involving hysterectomy, may be necessary in some cases but many centres now favour radiotherapy, a technique that has improved considerably in the last few years. New drugs and ways of administering drug treatment are also improving the treatment of cervical cancer.

The prevention of cervical cancer
Of course many cases could be avoided altogether if all women had regular smear tests, but the risk factors associated with the disease also suggest other ways of avoiding this condition. Women who have never had sexual intercourse do not get cancer of the cervix. But for women who don't want to opt for celibacy there are other ways of reducing the risk of developing

the disease. There does seem to be a link between certain viral infections and cervical cancer so trying to avoid coming in contact with these viruses by limiting the number of sexual partners and by using barrier methods of contraception would seem to be an effective way of avoiding cervical cancer *and* other infections. Some recent research has even suggested that condoms may help to bring about regression of cervical cancer. In one study, when condoms were used by the partners of women diagnosed as suffering from cervical cancer, the condition improved, possibly because the cervix had a chance to recover while protected from the cancer-causing agents in the men's semen. Avoiding intercourse in the early teens, when the cervix may still be developing, would also seem to be a good idea.

Cancer of the ovary

Cancer of the ovary is more common than cervical cancer. Most deaths occur in women in their fifties and sixties but the condition is not often diagnosed in the early stages as there are usually no symptoms, and it is difficult to treat. In the later stages possible symptoms include abdominal pain, abdominal swelling and abnormal vaginal bleeding.

Lumps on the ovary can now be observed using ultrasound equipment. However, experts disagree on the value of mass screening for ovarian abnormalities using ultrasound and on how often it should be carried out. Screening for ovarian cancer would be very expensive and it also picks up many abnormalities that have to be investigated further, often by surgical techniques, which then turn out to be benign. Many women would be subjected to unnecessary surgery, so widespread screening is unlikely to be recommended until screening tests become more sensitive.

It is possible to have an ultrasound scan of the ovaries at some of the private health screening centres but this is probably only worthwhile for women who have a close relative who has had cancer of the ovary.

Treatment
If a physical examination by a doctor together with ultrasound indicate that there may be a tumour on the ovary, an exploratory operation will probably be carried out to make a more accurate diagnosis. If the tumour is found to be malignant, treatment may involve removal of the uterus, the ovaries and any other affected tissues. Radiotherapy or drug treatment may also be given. But young women who have not yet had children may be successfully treated with less drastic surgery, particularly if tumours are diagnosed in the early stages, thus preserving their fertility. The chances of complete cure are not so good if the tumour is not diagnosed until it is well advanced. Overall, only thirty per cent of women with cancer of the ovary survive for more than five years.

Cancer of the endometrium

Cancer of the endometrium, or lining of the womb, is nearly as common as cancer of the cervix. It usually affects postmenopausal women in their fifties or sixties, and is equally common in women who have had children and those who have not.

Hormonal factors seem to be involved in the development of endometrial cancer – women who were given oestrogen therapy to treat menopausal symptoms were found to be at increased risk of endometrial cancer until another hormone, progestogen, was given as well as oestrogen. Patients with endometrial cancer tend to be obese so eating healthily and staying slim may be one way of guarding against it.

Researchers are trying to develop tests to improve early detection of the disease but until suitable tests are available women are advised to go to their doctor as soon as they experience any unusual discharge, postmenopausal bleeding, or bleeding between periods.

Diagnosis will usually involve a D and C and the basic treatment is surgical, involving removal of the womb, the ovaries and some of the vagina. In addition, radiotherapy given before or after the operation can help to prevent recurrences, and progestogen treatment may be given to treat the disease if it spreads to other sites. There is a good chance of surviving more than five years if the disease is confined to the womb.

Cancer of the vulva

This is a rare condition, mainly affecting elderly women – an average gynaecologist may see only one or two cases each year. Suspicious patches of tissue in the vulval area can be investigated with biopsy and cancerous areas will be removed surgically. Survival rates are good provided the operation is carried out by a surgeon who has much experience of this condition and the cancer has not spread.

Cancer of the vagina

This is even more rare than cancer of the vulva and treatments are not very successful. Surgery can be difficult, and radiotherapy or drug therapy may be more appropriate.

14 *Sexually Transmitted Diseases and Infections*

by Antonia Rowlandson

Infections of the genital and urethral areas can be caught during or caused by sexual activity, but some problems such as thrush and cystitis can affect women who never have sex. Poor hygiene, diet, stress and chemical irritants such as strong detergents can all increase a woman's vulnerability to infection. Medical treatment for most infections is effective but there is much that women can do on a practical level to avoid these unpleasant problems and, in some cases, to t   it themselves.

Vaginal discharge

Normal discharge
Many women find that secretions from the vagina vary throughout the menstrual cycle. Around the time of ovulation in the middle of the cycle vaginal discharge tends to be clear, thin and elastic. But after ovulation the mucus becomes thick, white and sticky. These secretions are normal and help to keep the vagina moist and healthy. It is important for a woman to be aware of what her secretions are normally like so that she is able to detect any changes that may indicate that there is a problem needing a doctor's attention.

Problem discharge
Stress, anxiety, illness or taking the contraceptive pill may all affect the amount and type of discharge. This is nothing to worry about but if it becomes a nuisance or is uncomfortable then it is a good idea to ask the doctor for help. If the discharge is an unusual colour, smells strongly or if there is itching or burning there may be an infection of some sort.

Prevention of problem discharges and infection

The genital area is sensitive and vaginal deodorants and antiseptic creams or lotions should be avoided as they can upset the delicate balance of the vagina responsible for keeping it healthy. Washing underclothes in strong detergents may also cause problems.

It is important to keep the vulval area as clean as possible by wiping backwards after a bowel movement and by washing regularly, but excessive use of soap should be avoided. Some women also find that bath additives can cause irritation. During a period, sanitary pads and tampons should be changed regularly – and forgetting to remove the last tampon at the end of a period is a fairly common cause of an unpleasant discharge. Highly absorbent, 'super' tampons should only be used during very heavy periods and it is best not to use tampons when the flow of blood is light since they can dry up important vaginal secretions. In addition, infections will be less likely to take a hold if loose-fitting clothes and cotton pants are worn.

To avoid picking up infection from a sexual partner, a woman should never have sex with someone who appears to have symptoms such as sores or a rash in the genital area. Limiting contact with casual sexual partners and using condoms are also highly advisable since symptoms of infection are not always obvious.

Where to go for help

Some infections may be diagnosed when a woman goes to her doctor for a smear test or contraceptive advice. However, if she suspects that she has a problem at any other time, she should seek help as soon as possible – some of the more serious infections can cause complications such as infertility if not treated promptly. Many infections of the female genital tract are very common and easily treated, so a woman should not feel worried or embarrassed about consulting a doctor about them.

Treatment from the GP will be sufficient in most cases, but if a woman does not want to take her problem to her GP for any reason, an alternative is one of the Special (genito-urinary or VD) Clinics at her local hospital. It is not necessary to be referred to these clinics by a GP but an appointment may be necessary – details of your nearest clinic can be found by looking under Venereal Diseases in the telephone directory or phoning your local hospital for details.

One advantage of the Special Clinics is that they have excellent facilities for diagnosing particular infections and are able to offer the most up-to-date treatment. A woman should be prepared to answer questions about recent sexual partners, since the clinic may well want to trace contacts in order to stop the spread of infection. Of course, all patients are assured of complete confidentiality. Staff at Special Clinics should also be sympathetic, supportive and in some cases will be able to refer a patient to someone who can give further advice and support if more help is needed. But clinics tend to be rather busy – long waits for treatment are by no means uncommon – and many women may prefer to consult and get treatment from their own GP who knows them well, rather than from an anonymous, overcrowded hospital clinic.

AIDS (Acquired Immune Deficiency Syndrome)

The predominantly sexual disease AIDS has now been with us since the beginning of the decade. Although experts are working on both a vaccine and a cure, there is still none available. The nature of the virus is still largely unknown, but, by charting the statistics over the last six years, a pattern has emerged showing that an increasing percentage of those whose blood is found to be antibody positive to the HTLV-III virus go on to develop the full-blown disease. The percentage of those dying as a result of fully developed AIDS also continues to rise.

In the United States, and other parts of the world where the disease has taken hold, notably central Africa, the number of women with AIDS is very much higher than in Britain. In many African countries, where sanitation and hospital hygiene are inadequate, AIDS affects men and women equally. It has become clear therefore that women are at risk from the disease, that it is not just a 'gay plague' affecting homosexual men. All the sexual partners of people in the high-risk groups – homosexual and bisexual men, intravenous drug abusers (through sharing needles) and haemophiliacs (through contaminated blood products) – place themselves in the high-risk category.

The AIDS virus is known to be transmitted in blood and semen, and therefore intimate sexual contact with an infected person should be avoided as should contact with infected blood. There is no evidence that AIDS can be passed on during normal social contact, and although the HTLV-III virus can survive in saliva and tears it has never been passed on in this way. Technically, people do not die from AIDS. Instead they succumb to one of twenty-two opportunistic infections which are not normally fatal but which become lethal in a body that has no power to resist or to fight back due to the destruction of white blood cells by the HTLV-III virus. AIDS works by attacking the body's immune system.

The Terrence Higgins Trust advises all sufferers on all aspects of the disease and has published a special pamphlet called *Women and AIDS*. In particular they warn women at risk from AIDS about getting pregnant. This is because there is a very high risk of passing the virus on to the baby either before or at birth. Already several babies have been born with AIDS in Britain. Pregnancy also increases the chances of an AIDS-infected woman developing the full-blown disease herself. Female carriers should not breast-feed because there is a risk that the virus could be passed on to babies previously not infected. Some women may be thinking of using artificial insemination, in which case they should be extremely careful about where the donated sperm comes from. The main agencies who specialize in artificial insemination are now screening donors, but it is wise to double-check.

All the AIDS organizations give clear details on 'safer sex'. Not wishing to isolate sufferers, these guidelines show how it is possible to continue physical contact, without risking infection, even when living or sleeping with a victim. Sharing towels and crockery is not hazardous, although some people may want to take precautions. However, people should not share razors or toothbrushes since both these objects can come into contact with infected blood.

Fortunately, the AIDS virus is fragile and does not survive long outside the human body, so unless it is directly transferred, through body fluids, the chances of infection taking place later are severely restricted. It cannot, for example, survive exposure to heat or household bleach. By far the main risk to women is through heterosexual intercourse with a high-risk individual. The best precaution is to think about safer sex, and to consider

---

Safer sex for women

High-risk practices
- Vaginal intercourse without using a condom
- Anal intercourse (the rectum is easily damaged and bleeds)
- Any act that draws blood
- Any act that involves contact with urine/faeces
- Sharing sex toys e.g. vibrators
- Sex during menstruation if the partner has open cuts/grazes/sores

Medium-risk practices
- Oral sex by the woman on the man, even if he withdraws before ejaculating
- Oral sex between women if the active partner has bleeding gums/mouth ulcers
- Oral sex by the man on the woman if he has broken skin in or near his mouth
- Vaginal intercourse with a condom

No risk (providing there are no cuts/bruises/open sores)
- Stimulation using hands, either mutually or solo
- General body contact e.g. stroking, kissing
- Orgasm and ejaculation of semen onto partner's body away from orifices
- Sex toys – as long as they are not shared

Source of information: Terrence Higgins Trust

---

condoms as an essential form of protection with someone whose past sexual activities are either unknown or make them suspicious, for instance a declared bisexual.

It is hoped that trials with the drug AZT will continue to show promise in prolonging the lives of victims until a cure is found and that for everyone else, male and female, homosexual and heterosexual, an effective vaccine will soon be developed and made available to all.

Thrush (Candida or Monilia)

Thrush is a very common problem which many women suffer from at one time or another and is caused by a yeast-like fungus called *Candida albicans*. This organism is often present in the body, particularly in the bowel, but is usually kept under control so that it does not produce any symptoms. However, if for any reason the delicate balance of the body is upset, it can proliferate, often causing unpleasant symptoms. It can affect both adults and children and may occur in moist areas of the body such as skin folds, the mouth, respiratory tract and vagina.

Symptoms
When thrush multiplies in the vagina, it can produce a thick, white discharge, sometimes a bit like cottage cheese. There may also be very uncomfortable itching and burning in the vulva and vagina. This can make intercourse extremely uncomfortable and there may be pain on passing water. But symptoms can vary enormously. Some women find that thrush gives them a watery discharge. Others have intense burning and discomfort but no evidence of thrush; in these cases, stress or a chemical irritant such as a different washing powder may be the problem. The only way to be sure that thrush is responsible for symptoms is to have a vaginal examination and a swab taken for laboratory analysis.

What causes thrush?
Thrush can be sexually transmitted – many doctors also treat the partner of a woman who has thrush to prevent the infection being passed backwards and forwards, particularly if the woman suffers from repeated attacks. However, women who are not sexually active can get thrush, and it is probable that the majority of cases occur spontaneously and are not sexually transmitted. Damage or inflammation of the vulval or vaginal tissues, caused either during intercourse or by an irritant such as perfumed soap, can increase the chances of thrush getting a hold.

Antibiotics are also a common cause of thrush – they kill off the bacteria that normally keep the growth of thrush under control. This is one good reason for not taking antibiotics unless really necessary. However, there is one type of antibiotic with in-built antifungal medication, so it is worth asking your doctor. During pregnancy, thrush may occur due to increased levels of oestrogen, and some experts also think that the oestrogen content of many contraceptive pills makes women on the Pill more vulnerable to the infection. Reduced immunity during illness or when a woman is over-tired or under stress may also lead to an attack of thrush. Recurrent attacks may be a sign of diabetes, and people consuming a diet high in sugar and refined carbohydrates may be particularly susceptible.

Thrush thrives in warm, moist conditions so that wearing tight clothes, particularly those made of synthetic materials, may encourage the growth of the yeast.

Treatment

Treatment should not be given unless thrush has been diagnosed by taking a sample for testing. One-off attacks of thrush, perhaps during pregnancy or after a course of antibiotics, are usually successfully treated with anti-fungal pessaries which are inserted into the vagina at night. Antifungal cream is also applied to external areas to eliminate traces of thrush in the vulva and perineum, and the woman's sexual partner may be encouraged to use the cream as well.

Treating recurrent thrush

Some women suffer from recurrent attacks of thrush – the antifungal treatment works but the infection returns a few weeks later. This can be very upsetting, particularly because it can seriously affect a woman's sex life. She may find herself trapped in a vicious circle where symptoms make her more and more miserable, which in turn tends to make symptoms worse. It is often difficult to find out why a woman goes on having attacks of thrush but there is much she can do to get to the root of the problem. There is nearly always a reason why the body is unable to keep the yeast under control.

First of all, she must follow the preventive measures discussed above – that is, she should wear loose-fitting clothes and cotton pants, pay attention to hygiene, and make sure that there are no irritants such as deodorants that could be aggravating the condition. If lack of lubrication during sex could be the problem, then using a lubricating jelly, such as KY jelly, should help. Some women find that coming off the Pill improves symptoms, and, if stress or exhaustion could be the problem, some form of relaxation such as a new hobby, sport or yoga, and getting more rest, could help.

Some doctors treat recurrent thrush with tablets of an antifungal called Nystatin. This is meant to reduce the prevalence of thrush in the bowel, and thus prevent reinfection from this source. For some women, inserting live, natural yoghurt into the vagina, possibly with a tampon, every day for a few days can get rid of thrush. Bacteria in the yoghurt help to keep the Candida organism under control.

If none of these strategies work, the next step is to look at whether changes in diet could deal with the problem. Your GP may be able to help with this approach but many women find that an alternative practitioner, such as a naturopath, homeopath, herbalist or acupuncturist (who will also treat stress), may have greater expertise and knowledge of this area.

The Candida organism thrives on refined carbohydrates so it is important to cut down on sugar and starch. Foods containing yeasts and moulds are also thought to be a bad idea, so that means no sugar or refined cereals or products containing them, no bread made with yeast, cheese, alcohol or malted products. Some experts also recommend cutting out all milk products, vinegar, coffee, tea and smoked foods, and all fruits should perhaps be avoided for the first two weeks of the diet (they contain sugar and may also have some mould on them). This will be a difficult diet to follow and probably only those with severe thrush will have enough motivation to stick to it. It is important that there is adequate protein intake and a woman should only go on this diet with some sort of expert supervision.

Alternative treatment may also involve taking supplements including vitamins A, C and E and also zinc. The B vitamins, together with plenty of exercise and sleep, should help to boost the immune system. Some practitioners are also keen on supplements of *Lactobacillus acidophilus,* but these can be expensive and simply eating live natural yoghurt may be a cheaper alternative.

Trichomonas vaginalis ('trich')

This organism is probably spread by sexual intercourse and infection is most likely to occur just after a period, when the vagina is more alkaline. Symptoms include a frothy, greenish or yellow discharge which may have an unpleasant smell and which can cause itching and irritation. But men often have no symptoms. Treatment is straightforward involving the prescription of an antibiotic, usually Metronidazole. This antibiotic can cause stomach upsets, so alcohol should be avoided when undergoing this form of treatment. It is also important to remember that the infection will recur if *all* sexual partners have not been treated.

Chlamydia trachomatis

This sexually transmitted disease has been receiving more attention recently because it seems to be on the increase and may cause problems such as pelvic infection, ectopic pregnancy and even infertility. Possible symptoms are pelvic pain, bleeding between periods or after intercourse, or urinary problems. However, in many cases it produces no symptoms in men or women – damage to the fallopian tubes can be silent, with minimal symptoms.

There is a simple test for the disease but many doctors think that this diagnosis is not considered often enough, and that routine screening should be carried out on women with abnormal vaginal discharge and pelvic pain and also those at increased risk of becoming infected, for example those with many sexual partners or whose partners are promiscuous. The infection can be passed on to the babies of mothers who are infected and may also be linked to premature birth and stillbirth. Widespread screening at antenatal, family planning and pregnancy advice centres would be a good idea, particularly now that cheaper diagnostic tests are becoming available. Treatment with antibiotics is usually very effective.

Genital warts

In women, these sexually transmitted warts may grow in the vagina, in the vulva area or on the perineum. They may be anything from small lumps to large cauliflower-like growths. The wart virus is thought to play a part in the development of cervical cancer, so a woman should try to avoid having intercourse with someone who has warts until they have been treated, and if she thinks she has warts herself she should seek treatment as soon as possible. Warts can easily be treated.

Pubic lice ('crabs')

These parasites can be picked up during sexual contact and sometimes from beds, clothes and towels. They live in pubic hair, feeding on human blood, and have small white eggs. They may cause itching but are easily treated with a special lotion available from a clinic or doctor.

Genital herpes

Herpes is a virus nearly always transmitted by sexual intercourse. Its incidence seems to be on the increase and much alarming publicity about herpes has resulted in many people being very frightened of the disease – a diagnosis of herpes can spark off panic and despair. Although there is so far no cure for herpes, there is much that can be done to control symptoms and it certainly does not put an end to sex.

The herpes virus
There are many different types of herpes virus including *herpes zoster,* which is usually known as shingles and is not linked to genital herpes. Genital herpes is caused by the herpes simplex virus of which there are two types. Type 1 usually causes 'cold sores' around the mouth and nose, whereas type 2 is more often responsible for sores in the genital and anal region. But doctors are now seeing an increasing incidence of genital herpes caused by type 1, so it is very important that people with cold sores do not have oral sex as they may pass the infection to their partner's genital area. Herpes can also be caught during genital contact when one partner has sores, so it is important not to have sex with anyone with these symptoms.

Symptoms
The main symptoms are small blisters in the genital and possibly anal areas. A woman may also have blisters on her cervix, particularly in a first attack. The blisters eventually burst leaving painful ulcers which may take two or three weeks to heal. The virus then enters the nerve fibres and travels up the nerve to the nerve root, where it is thought to lie dormant for the rest of a person's life. Some people have just one attack and never have any more problems, but in others the virus is reactivated from time to time so that they go on to have several further attacks over several years. In general the first attack is the worst, and it may be accompanied by flu-like symptoms such as feverishness. Warning signs of an attack include tingling or pain in the area where a blister may appear, headache, pains in the legs and the glands in the groin, and a general unwell feeling.

Diagnosis and treatment
It is most important to go to the doctor straightaway if herpes, or any other infection, is suspected. GPs should be able to diagnose and treat herpes but a Special Clinic at a hospital that specializes in sexually transmitted diseases will have more facilities for making an accurate diagnosis and should help with any queries or worries that patients may have about the disease. If herpes is diagnosed there is no need to panic. Firstly, this may be the one and only attack, and, secondly, if there are further attacks, there are lots of ways of reducing the discomfort of symptoms. Besides, herpes does not affect a person all the time – for much of the time a woman can forget all about it, though she should tell sexual partners that she has the infection, explaining that the risk of passing it on between attacks is very small. Sex should always be avoided before, during and just after an attack.

The doctor may prescribe a new drug called acyclovir (Zovirax) which may reduce the severity of the infection if given during the first attack. Sometimes the very small number of people who suffer from severe, recurrent attacks are given regular treatment with acyclovir. However, doctors are reluctant to prescribe extensive treatment with this drug, partly because they are unsure of its long-term effects on the body and partly because the body may become resistant to it. It is also expensive. There is an acyclovir cream which can help to prevent or reduce the severity of an imminent attack.

Treating the symptoms
An analgesic such as aspirin will help if the pain is severe, and some people find that adding a handful of salt to a warm bath soothes the sores. Applying an ice pack or dabbing on witch hazel or surgical spirit may also help. The flow of urine over open sores may be extremely painful and passing urine in a warm bath may be a solution to this problem. The sores

will heal more quickly if irritation is kept to a minimum – for example, by wearing loose-fitting clothes – and if they are left exposed to the air as much as possible. There are also several homeopathic remedies for herpes and its symptoms.

Reducing the frequency of attacks

Keeping the genital area clean, avoiding anything that may irritate the skin and wearing loose-fitting clothes and cotton pants should help to reduce the frequency of attacks. It is possible that friction during sexual intercourse may sometimes trigger an attack so it is important to make sure that there is always sufficient lubrication during intercourse, perhaps by using a lubricating jelly. Excessive sunbathing and use of sunbeds are also thought to encourage reactivation of the virus.

Many people find that attacks tend to occur when they are tired or after a period of stress. So avoiding stress whenever possible, taking lots of exercise, getting adequate rest and perhaps taking up some stress-reducing activity, such as yoga, may reduce the number of attacks in susceptible people. Some women find that attacks are more likely at certain times in their menstrual cycle, so getting as much rest as possible during these vulnerable times may help.

It is also most important to stay on top of the problem and to keep it in perspective. Negative thoughts will do nothing to boost resistance to the infection or help a person cope with the symptoms. Eventually most people come to terms with herpes, work out the best way of dealing with their symptoms and soon feel able to resume normal relationships. However, some people may need extra help. The Herpes Association gives support and information and has details of self-help groups (see Useful Addresses).

Nutrition and herpes

Another approach to recurrent attacks of herpes is to try to increase the body's resistance to infection through diet. This involves eating a healthy diet and possibly taking vitamin and mineral supplements such as vitamins A, B6, B12, C, E and zinc. Some alternative practitioners are now also recommending an amino acid supplement called lysine and a diet that is high in lysine-containing foods and low in another amino acid called arginine. It is thought that lysine inhibits the growth of the herpes virus whereas arginine encourages it. Lysine-rich foods include chicken, fish, beef, lamb, milk, cheese, brewer's yeast and mung bean sprouts. Arginine-rich foods to be avoided include chocolate, gelatin, nuts, oats, wholewheat and white flour, soya beans and wheatgerm.

There is no firm evidence that this approach works but it is one more line of attack that people who suffer repeatedly from herpes might like to try, and it is unlikely to do any harm.

Avoiding the spread of infection

People with cold sores should refrain from kissing and oral sex, and those with genital sores should avoid genital contact with a partner before, during and for a few days after an attack. It is unlikely that herpes can be passed on from towels and flannels but it is nevertheless sensible for someone with herpes not to share these with other people. To avoid spreading infection to other parts of the body, the hands should always be washed thoroughly with soap and water after touching sores. In particular, fingers should not be put anywhere near the eyes after contact with the infected area(s).

Herpes and cervical cancer

Some of the alarming publicity about herpes arose from suggestions that women who have genital herpes are more likely to develop cervical cancer. However, there is no firm evidence of a direct link between herpes and cancer of the cervix. To be on the safe side, women who have herpes are recommended to have an annual cervical smear (see Chapter 13).

Herpes and pregnancy

Herpes infections in babies are very rare, but if a woman has open sores when she gives birth the baby may become infected with the virus. A woman who has herpes should inform her obstetrician and should have regular checks for signs of an attack as the time of the birth approaches. If there is evidence of an attack a Caesarean delivery may be recommended.

A first attack of herpes during early pregnancy may lead to miscarriage, but recurrent attacks will not affect the baby in the womb. It is important to make sure that a baby does not come in close physical contact with someone who has cold sores, but it is very unlikely that a baby will be infected by a parent who has genital sores.

Gonorrhoea (the 'clap')

This is a sexually transmitted disease in which symptoms start a few days after vaginal or anal intercourse or oral sex with an infected person. A woman may notice that it is painful to urinate, she may have a discharge and possibly fever, abdominal pain and painful joints. In men there may be pain when urinating and a yellowish discharge from the penis. Sometimes symptoms are so mild that a woman does not know that she has the infection. Therefore, women who have many sexual partners or whose partners have other sexual relationships should have regular checks for the disease. If it is not treated with antibiotics the infection may spread, causing pelvic infection and infertility. Long-term effects of untreated gonorrhoea include permanent damage to joints.

Syphilis (the 'pox')

This is also a sexually transmitted disease but it is not very common today. The first symptom is usually a painless sore which appears in the genital region several weeks after infection. In the next stage there may be a rash, mouth sores, flu-like symptoms and a general feeling of being unwell. In the third stage several months or years later there may be considerable damage to the heart and blood vessels, to the central nervous system and to the eyes. The disease can be successfully treated with antibiotics if diagnosed by blood tests in the early stages. Treatment in the later stages may eliminate the infection but may not prevent the long-term complications.

Pelvic inflammatory disease

Symptoms include lower abdominal pain and fever. There may be inflammation of the fallopian tubes, of the ovaries or of the peritoneum. Pelvic inflammatory disease may follow an infection associated with childbirth or abortion. Infections such as gonorrhoea and chlamydia may also give rise to this disease. IUDs (or coils) are also thought to increase the risk. Pelvic inflammatory disease should always be treated promptly as it can be serious and may damage tubes and ovaries and cause infertility.

Non-specific urethritis (NSU)

This infection of the urethra is now one of the most common sexually transmitted diseases, particularly in men. A man may have a burning sensation when he passes water and may notice a discharge from the penis. Women get NSU but very few have symptoms. Chlamydia is often the cause of NSU and a course of antibiotics for the patient and his or her sexual partner(s) should clear up the infection.

Cystitis

This is not a sexually transmitted disease but sexual intercourse is one of the many causes. Cystitis means inflammation of the bladder, and can be caused by various germs. Symptoms include a burning sensation when passing water and a frequent need to pass water. Urine may be strong-smelling, cloudy and even bloodstained. There may also be fever and pain in the lower back or abdomen. Cystitis is more common in women than in men, mainly because a woman's anatomy makes her more vulnerable to this infection: the urethra is short and the vagina, anus and urethral opening are all very close together so that bacteria from the bowel can easily get into the urethra.

The causes of cystitis include bruising of the urethra during intercourse, particularly if the woman is not sufficiently aroused or if she is using a contraceptive cap that does not fit properly. Chemical irritants such as bath additives, washing powders or spermicidal creams may also bring on an attack. Cystitis also tends to be more common during pregnancy and in menopausal women – the ovaries begin to release less oestrogen during the menopause and this makes the bladder and urethra more vulnerable to infection.

A single attack of cystitis may be effectively treated with antibiotics but some women suffer from recurrent bouts. Repeated courses of antibiotics are not a good idea because they tend to wipe out the 'good' bacteria so that other poblems such as thrush can get a hold, and also because resistance to some antibiotics can build up if they are used too often.

How to avoid repeated attacks of cystitis

It is important to maintain good hygiene: always wipe carefully from front to back after a bowel movement. The genital region should be washed with water or unperfumed soap twice a day and washing before intercourse (both partners) is also advisable. If there is a lack of lubrication during intercourse, a lubricating jelly may be the answer. A woman should never have intercourse unless she feels like it and is sufficiently aroused. She should avoid having intercourse with anyone who has an infection and should pass water immediately after intercourse. Some women find that pouring cold water over bruised tissues also helps. Wearing loose-fitting clothes and cotton underwear should help to prevent infection. Certain foods and drinks such as tea, coffee, alcohol and spicy foods may also trigger an attack. It is important to keep the urine dilute by drinking plenty of fluids (water, diluted fruit juices or weak herbal teas) and to pass water often, making sure that the bladder is emptied each time. Anxiety, depression or worries about sexual relationships may also make a woman more vulnerable to cystitis and psychotherapy or counselling may be necessary to sort out these problems.

At the first sign of an attack of cystitis a woman should drink as much water as possible and keep emptying the bladder. This should help to flush out the infection. If an attack still occurs, she should carry on drinking as much as possible, take aspirin to relieve pain and hold a hot water bottle against the back or abdomen if they are uncomfortable. Drinking a teaspoon of bicarbonate of soda dissolved in a little water should also help to reduce the burning sensations associated with passing water. If symptoms of acute cystitis continue the woman should see her doctor.

15 *Sexual Violence*

by Patricia Gray

Sadly, sexual violence is a reality for women throughout their lives, and rape itself is a subject shrouded in half truths, fear and ignorance. The aim of this chapter is to give women a clearer understanding of rape and of the options open to them if they suffer a rape attack. It also gives basic information on how to take positive steps towards prevention.

Legal definitions and sentencing

Rape

Rape in *legal* terms is to have sexual intercourse with a woman knowing that she has not consented, or reckless as to whether or not she has consented. If a woman is attacked but not penetrated vaginally, it is *not* classed as rape even if acts such as oral or anal penetration are committed. It is felt by many women's organizations that any form of penetration should be regarded as rape. The main reason for this is that many other forms of penetration, with the exception of anal penetration, are classified as indecent assault, and this is seen in law as a 'lesser' crime and so carries a lighter penalty. Penetration of the vagina by the penis was originally deemed more serious because it can result in pregnancy.

Rape carries a maximum life sentence, though this sentence is very rarely enforced in the courts. The average sentence for men found guilty of rape has been two to three years, though longer sentences have been given to men found guilty of more than one rape and to men who have used excessive violence during a rape attack. In 1986 Lord Chief Justice Lane recommended that there should be a starting base of five years for rapists. He also recommended that there should be a higher starting base of eight years if there is more than one man involved (gang rape) or if the victim is either very young or very old.

Indecent assault

Oral penetration or penetration of the vagina by fingers or foreign objects fall into the category of indecent assault. Indecent assault carries a maximum of ten years' imprisonment. However, as with rape the maximum sentence is rarely enforced.

Buggery
This is the legal term for anal penetration. Buggery is considered as serious an offence as rape and also carries a maximum life sentence.

Rape and marriage
Legally a man cannot be charged with raping his wife. The reason for this is that when a woman marries a man she consents to sexual relations with him. A man can only be convicted of rape if it is proved that the woman did not consent. The law sees marriage as a consenting partnership and only if the wife has a legal separation does it recognize that consent has been withdrawn. Many women's organizations are lobbying for a change in the law, but, at present, a wife cannot bring a charge of rape against her husband. However, buggery is illegal whether the woman is married or not, and a man can be charged with buggery against his wife. Rightly or wrongly, many women feel that buggery is seen in the law to be as serious as rape only because it can be committed against men as well as women.

The reality of rape

Up until the last ten years or so rape remained a taboo subject rarely spoken about. Ignorance about what rape meant to the women who had experienced such an attack was common. Knowledge of the subject was gleaned primarily from newspaper reports and hushed whisperings. Rape was not considered to be widespread, or to have lasting effects on its victims. For the women who had experienced rape there was little support or understanding. Even worse was the fact that society did not even recognize that they may *need* understanding or support after an attack.

Although people's feelings about the women's movement are generally mixed, it has done a lot to change society's attitudes towards rape. It is responsible for the establishment of Rape Crisis Centres all over the British Isles, and these centres have contributed greatly to the public's awareness of the needs of the raped woman. The police have improved their treatment of women who allege rape and small changes in the law have resulted from campaigns by concerned women. However, one of the most important steps taken by Rape Crisis Centres has been to tear down the myths surrounding rape. These myths have to a large extent contributed to the trauma experienced by a woman after the attack. They have certainly been responsible for her feelings of guilt and shame.

There are many half truths and myths about rape, and these tend to abound in three areas: the theory that rape is motivated by the need for sex, the kind of woman who is raped and the man who rapes. It is important to discuss these areas first in order to understand the true nature of rape and to comprehend the numerous problems many victims encounter.

Sex or power

Rape has always been considered to be a sexually motivated crime. The man who rapes is often seen as a kind of maniac who has no control over his sexual drive. This belief is widely held and even today continues to be perpetuated through the media; it is not unusual to open a newspaper and read headlines such as 'Sex Fiend Rapes Again'. Women working with raped women have unearthed a great deal of evidence strongly suggesting that the motivation behind rape is not sex but power. This is also supported by the findings of people who work with the men who rape. Indeed, over the past few years there have been films and documentaries made with and about rapists, where the men have actually described their acts as power-based.

The rapist uses a very effective weapon – fear. Women who have been raped very often view the experience as life-threatening. Whether the rapist uses violence or the threat of violence, the woman is made to believe that she has no choice other than to submit to her attacker's wishes. It has become clear that an act of rape rarely consists of just the act of sexual intercourse. More often than not the woman is subjected to other acts, such as oral or anal penetration. She will be degraded and humiliated verbally and psychologically, she will be made to understand that she has no control over the situation. Once the connection between rape and power is understood, it becomes easier to see why women fall victim to this crime. Women are generally smaller physically than men but, more importantly, women are conditioned to be passive. They are not encouraged by society to be aggressive, indeed aggressiveness in women is seen as unfeminine or undesirable. Hence women who find themselves in a rape situation may be unable to defend themselves. The power theory also makes it clear why old women and children fall prey to sexual offenders – they can be easily overpowered and thus represent easy victims.

The 'bad' woman syndrome

For years it was a commonly held belief that only 'bad' women – prostitutes or the promiscuous – were the target of rapists. However, it has been the experience of women working in the field of rape that being what society terms a 'respectable woman' does not prevent rape from happening. Women of all classes, nationalities and ages are being attacked in this way. Many women feel too ashamed to go to the police, they fear that they will not be believed and that they will be labelled as 'bad' women. The sexuality of women is often questioned by society; ever since the story of Adam and Eve, women have been blamed for tempting men from the path of righteousness. The 'bad' woman myth has been used as a way of exonerating the rapist and laying the blame at the victim's door: the woman tempted the rapist beyond endurance, she provoked him and so therefore not only is she responsible for the rape but she deserved it. Again, the act of rape is being viewed sexually, the emphasis firmly rooted in the sexuality of women. Once rape is viewed from a power angle this myth is exploded.

Society constantly feeds a distorted view of rape, mainly through magazines, books, films and television. The idea that leads people to believe that it is harmless and enjoyable is a dangerous fantasy. Rape is often depicted as the forcible (not violent) taking of a woman, by someone she has chosen and more often than not given consent to. Thus in many books churned out for women there is a strong message that rape is simply rough sex, that it is not unpleasant. In reality a woman has no choice. The rape is brutal either physically or psychologically or both, and her consent is never sought. It is because of this confusion of fantasy/reality that we frequently hear women say things like 'I should be so lucky' when the subject of rape is mentioned.

Rape is frequently the subject of jokes. It is not funny nor is it enjoyable. Rape is a reality, one that pervades our entire society. In numerous ways the responsibility for rape is shifted to the victim herself. If a woman is dressed 'provocatively', or if she is out late at night or if she has no male protector, she is somehow seen as, and indeed is often accused of, contributing to her attack. Many men are attacked, beaten or mugged. However, there is one major difference: whether they were out late at night or how they were dressed or the circumstances of the attack are never seen as contributory factors and used against them.

Man or monster

The typical response to the man who rapes is that he must be a monster, a maniac, set apart from other men. This monster/maniac loiters in dark alleys, or on waste ground, ready to pounce on his unsuspecting victim. It is true that some rapists have been classified by psychiatrists as psychopathic, though this has generally been the notorious multiple rapist who comes to the attention of the authorities, the press and the public. This type of rapist has brought about the myth of the monster/maniac.

Many women have been raped but not by the man described above; they have been attacked by men they know and, more importantly, men they would never have thought to be monsters or maniacs. The myth has led women to a false belief, and so made them vulnerable to such attacks. Another fact which has come to light over recent years is that rape attacks do not always happen late at night, out in the open. A lot of women have been raped in their own homes or the home of their attacker or in other familiar surroundings – they are not attacked just by strangers. The rapist is first and foremost a coward, a man who needs to feel powerful. In order to be powerful he must choose a victim he considers to be weaker than himself and who does not suspect that he will attack them. For the rapist, surprise and fear are his weapons. For the woman, disbelief, shock and horror all play a vital part in her attack.

Where to go for help and advice

Rape Crisis Centres have been set up in order to meet the specific needs of women and young girls who have suffered a rape attack. In the past women who had been raped often kept it to themselves or turned to people such as doctors or the clergy. Unfortunately, their needs were frequently misunderstood and not catered for. It takes specialist knowledge and counselling to help women come to terms with the experience of rape.

If you or someone close to you has been raped or sexually assaulted, you can contact your nearest centre for help and advice (you will find the number of the centre near you in your local telephone directory). If you phone a Rape Crisis Centre you will not be asked for any personal details (unless you wish to give them) – all centres are confidential. Their counsellors will listen to you, and give you information and/or advice on any problems you are encountering. All centres are non-judgemental; you will not be forced into any decisions or courses of action that you do not wish to take. Any decisions you make will be fully supported. If you wish to report your attack, it is important to know what to expect.

The police

If you are going to report, then do so as early as possible, as any delay may destroy vital forensic evidence that will corroborate your allegation. Sexual offences are the only group in which a jury is warned that it is dangerous to convict on the evidence of the complainant alone. This means that the woman's word is not evidence enough, she must have concrete corroboration. Since there are usually no witnesses, forensic evidence is normally the only form of corroboration.

When you go to the police to make an allegation, you can if you wish take a friend or relative. Alternatively, if you contact your nearest Rape Crisis Centre a counsellor will accompany you to the police station. Firstly, you will be taken to an interview room where it is normal for a police woman to interview you and take your statement. This first statement will be long and detailed. You will be expected to recall all the events as they happened and you must describe every sexual act you were forced to endure. This can be very distressing, but remember that the police will expect you to describe these events fully, so take your time and do not allow yourself to feel under pressure (if you do feel distressed and pressurized, make this clear to the police woman). The police have come a long way in trying to understand how women feel after a rape attack. They usually incorporate special training in their course on rape, and they now allow breaks in the statement-taking to enable women to calm down and feel more comfortable. Try not to rush your statement as you may forget important facts, or

Left: The Vitullo Evidence Kit which, because of its success in the United States, may soon be introduced in Britain. *Below:* One of the new informal rape interview rooms, where a police woman takes the statement of attacked women.

become confused. If you go to the police station wearing the clothes in which you were attacked, the police will want to keep them to send to the laboratory. If you have changed your clothes the police will still want them, so they will go to your home to collect them. A lot of police stations now have tracksuits for women to change into; otherwise, police officers will go to a woman's home to collect some of her own clothes for her to wear. They need the clothing because it probably holds forensic evidence that will be important to the case.

Forensic evidence
While you are giving your statement, other police officers will be contacting a forensic surgeon to come and examine you. This may take some time, as police forensic surgeons are also general practitioners. However, since it is

now understood that the examination can cause great distress there is every effort made to get the examination completed quickly.

The examination will consist of you giving a brief statement about the rape to the surgeon. This is so that the surgeon can look for specific signs of bruising and other violence, which will verify your statement. During the examination vaginal swabs, oral swabs, and samples of pubic and head hair will be taken. You will be required to give a blood sample and also have your nails examined and cleaned. All of these samples and swabs are then sent to laboratories for analysis. The surgeon will also do a thorough check of your body for bruising and cuts. If you want to object to being examined by a male surgeon, or are in any way disturbed by the fact that he is male, you should express your feelings, as there *may* be the possibility of a female surgeon performing the examination. If there are visible signs of injury, these are generally photographed by a police photographer and produced in court as testimony to your injuries.

The statement and the examination last somewhere between four to ten hours and you are then allowed to go home. If you want to take a copy of your statement home with you for reference in court, it is your right to ask and be given one. You may be visited in the following week(s) by the officer in charge of your case – this is usually to check your statement. If they are able to detain and arrest your attacker and if there is positive forensic evidence, the crown prosecutor will decide whether there is a strong enough case to answer. A rape trial can take six months and upwards to come to court.

Remember if you decide to report

- Do not wash yourself
- Do not destroy or wash clothing
- Do not drink anything (mouth swabs)
- Do not delay in reporting
- Do not disturb any items within the room or house (if the attack takes place under these circumstances)

The courts

Rape is classed as a serious crime and will eventually come to trial in a crown court. However, the accused man will appear first in a magistrates court. At the hearing in the magisrates court it will be decided whether the man is to be kept in custody or allowed bail until the trial. The police generally object to bail; but it is the magistrates who take the final decision. The magistrates will always refer the case to a crown court, where the case will be heard by a judge and jury if the accused man pleads not guilty. If he pleads guilty, the defendant will appear before a judge for sentencing and the woman will not be required to give evidence.

The woman who has been raped is notified by post, normally two weeks before the trial, though sometimes this can vary, or she may be contacted by the police and told the date of the trial. When she is given notification of the trial date she will also be told at which court the trial will be held and the time to attend. The prosecution barrister may want a brief word with the woman and this will be the first time she meets him/her. She is not allowed to talk to other witnesses and must wait until she is called to give her evidence – she is always the first witness to be called.

Usually, but not always, once she has given evidence the woman is released by the judge and allowed to leave the courtroom. Since the raped woman is the prime witness to the prosecution's case, she may be in the witness box all day and, if needed, the following day as well. She will be cross-examined by the defence, and the accused man will be present in the court while she gives evidence. Most women who make an allegation of rape that comes to trial often feel that they are the one on trial – and in effect they are. Rape is often just a matter of one person's word against another's, and this is why rape trials are often a humiliating and devastating experience for the woman.

Criminal Injuries Compensation Scheme

If you have been the victim of a crime of violence, you can apply to the Criminal Injuries Compensation Scheme. If you do, it must be done within the three years following the attack. If you make your application after the three-year limit, you must give a very good reason. Rape victims can apply for compensation even if the attacker is not apprehended. They can also apply if they go to court and their attacker is found not guilty. The Criminal Injuries Board (see Useful Addresses) will grant compensation, as long as they are satisfied that the injuries were incurred due to an act of violent crime. They will make enquiries of the police, hospitals and doctors to verify the extent of your injuries (physical and psychological), and, if they are satisfied, one member of the board will grant compensation. You will probably have to wait a while before you receive the award. Most Rape Crisis Centres carry the compensation forms, and they will also help you to complete them. You can also write direct to the board, who will forward you the forms with explanatory notes.

Pregnancy and sexually transmitted diseases

If you are raped, whether or not you decide to report your attack, there are two very important possibilities to be considered – pregnancy and sexually transmitted diseases. Some women who have been raped and had a forensic examination have thought that this included checks for pregnancy and sexually transmitted diseases. This is not so. The forensic test is done only to find evidence to corroborate your allegation. It is important for a woman who has been raped to have separate checks for pregnancy and infection, and to bear in mind the effectiveness of morning-after contraception in reducing the risk of pregnancy (see Chapter 5).

Reactions of family and friends

Family and friends usually have little understanding of rape, and they may experience feelings of anger, hopelessness, frustration and guilt. In many families there may even be doubt about the woman's allegation – especially if the accused is someone the family knew and trusted. Although most families want to help, they can (without realizing it) cause a lot of pain and anguish by trying to push the woman into the course of action *they* think she should take. For example, they may want her to report/not report without considering her feelings. It is sometimes helpful if other members of the family also receive help and have the opportunity to express their fears and anger. Rape Crisis Centres can offer advice and some counsel family members if they have the resources. If necessary, they can refer you to alternative organizations.

Avoiding rape situations

On a practical level there are a number of precautions that can be taken to avoid an attack:

- If you are out at night always stay in well-lit areas
- Always carry door keys in your hand so that you are ready to open your door
- If you drive, always check the back seat of your car before getting in
- Ask for and check the identification of any official callers to your home
- Fit a spy-hole and a chain to your front door; keep the chain on when first answering the door

Another very good precaution is to learn self-defence, since this can be used to protect yourself from strangers *and* people you know. There are many self-defence courses specifically designed for women; they teach basic moves that women of *all* ages can learn. These courses are usually held in leisure centres or schools and colleges, so phone around and check where there are courses near you. The police also offer some self-defence classes, so check with your local station. Women who have been raped do suffer terrible trauma, but with the right help they can come through the experience with a stronger and more positive attitude to themselves as women.

16 *The Health Professionals*

Visiting the GP, specialist or hospital can for some people be a frustrating and even distressing experience. The question of receiving information often lies at the root of this problem: either the doctor does not seem sympathetic and the patient is too afraid to ask about the diagnosis and the treatment being prescribed, or the patient does not have sufficient knowledge of common health problems to be able to ask the right questions. After all, in order to receive treatment at all you must first be able to describe your symptoms adequately. However, improved communication between doctor and patient will not automatically remove all the problems that exist in the health services, but increased health education and knowing your rights will certainly help.

Your relationship with your doctor

It is of vital importance to have an open and trusting relationship with your doctor. However, this may not be as simple as it sounds. Your relationship with your doctor may be affected by a number of pressures and, as a consequence, a lot is left up to the individual patient to make sure that she gets the most out of her consultation. She can, for example, make things easier by planning in advance what she is going to say – this will also help ensure that no important details are missed out. Similarly, on a practical level a patient should find out as much as possible about the GP, the practice and the services it offers before registering since this may avoid potential dissatisfactions.

Most people are happy with their doctor but if you feel that your relationship is not as harmonious as perhaps it should be, first ask yourself why. If you are in the habit of making frequent appointments for minor ailments (a cold, for example) or calling your doctor out at night when it would in fact have been possible for you to visit the surgery, you are almost guaranteed to provoke an unsympathetic response. If, however, you do not place these unnecessary demands on your doctor and still feel that you are being treated in an off-hand, dismissive way you could try being more assertive. Make it clear that you demand the proper service to which you are entitled, that you will not be deterred by the fact that the doctor seems to be in a hurry and the waiting-room is full. After all, you will probably only see him/her three or four times a year if you are in reasonably good health, so it is not too much to ask to be given time and due consideration. Legally, it is your most important right as a patient to be treated with reasonable care and skill.

If you have a male doctor, your relationship with him may be affected if you are shy or embarrassed, particularly where matters of your sexual health are concerned. Similarly, certain religious and ethnic traditions make it difficult for some women to talk about problems of an intimate nature, especially to men. If you do not find it comfortable talking to a male doctor you should let him know. He will usually be sympathetic and advise you either to see a woman doctor in the same practice or, if there is none, to change to another practice where you will be able see a woman practitioner.

Receiving information

Once a diagnosis of your condition or illness has been made, you should feel able to discuss the causes and possible methods of treatment with your doctor. He or she should explain clearly how the treatment works, whether there are any risks involved and what alternatives are available if the one prescribed fails to sort things out. Fortunately, doctors of both sexes are becoming increasingly aware of women's problems such as pre-menstrual syndrome and the menopause and are more sympathetic in their approach. Not only are they more alert to symptoms and their psychological as well as physical effects but they are also able to suggest a wider range of treatments, both conventional and alternative. Women can also actively contribute to this greater awareness by their own increased knowledge, ability to recognize symptoms and participation in decision making. The latter is particularly crucial in cases where there is doubt about which form of treatment is most effective; for example, in the treatment of breast cancer, lumpectomy may be just as successful as mastectomy. The final decision should be left to the woman.

Frank discussion with your doctor is obviously the best way of getting information, but there are occasions when a doctor may deem it in the patient's interests to withhold information and the results of tests, for example when these reveal an illness that cannot be treated. It could be argued that a patient has the moral right to know, but she does not have the legal right – medical records belong to the Secretary of State for Health. If you ask, you will usually be told what is written on your records, but a doctor cannot be made to give out this information. In hospitals, lack of information is more often than not the result of a failure to communicate, and asking politely but firmly for details of tests and treatment, and discussing them with your doctor or a nurse will often sort things out.

Getting a second opinion

It is only possible to see a specialist under the NHS by first obtaining a referral note from your own GP. This should not be a problem and the majority of people who ask their doctor for a second opinion are referred to a relevant specialist. However, since the specialist is usually a hospital consultant, GPs are loath to refer patients when they do not consider it necessary; they have an obligation not to waste the specialist's time.

If you are dissatisfied with your treatment and your GP fails to explain why he does not consider referral to a specialist necessary, you have a number of options. If the practice is large, you might consider seeing one of the other GPs; this may be practical in terms of travelling to the same location, but may cause problems with your previous doctor. Alternatively, you could change to a new GP in a different practice who may either offer a different form of treatment, an alternative medicine for example, or be willing to refer you to a specialist. It would be very unusual to find a specialist who did not require a referral letter from the patient's doctor. Often, talking things over with your GP is the best way of reaching a solution that provides the most effective care for the patient as well as maintaining an open and trusting relationship with your doctor.

If you have been referred to a specialist by your GP and he or she recommends major surgery, such as hysterectomy or mastectomy, you might want to get the opinion of a second specialist before going ahead. The first specialist should be able to help, but if this proves difficult for whatever reason you can always go back to your GP for a referral. You will also want to talk things over with your partner or close relatives, so try not to be in too much of a hurry, though excessive delay should, of course, be avoided.

Changing your doctor

There can be many reasons for wanting to change your doctor. A change of address is perhaps the most obvious: if you move to another area or town you are strongly advised to register with a new GP as soon as possible since otherwise you will not be eligible to receive treatment unless it is an emergency. If you were even mildly dissatisfied with your previous GP, it would be unwise to choose a new one at random. Make sure that the practice offers the services you require before you register: check, for example, whether you can see a woman doctor if this is what you would prefer. Other services you might consider an advantage, or even essential, include well-woman clinic services, preconceptual care, antenatal clinics, maternity services, immunizations for small children, cervical smears and contraceptive advice. Information can be obtained from the Family Practitioner Committee (FPC) Medical List, The Medical Directory, local chemists and, perhaps most valuable of all, the opinions of neighbours!

It is not necessary to give a reason for wanting to change your doctor and the procedure is quite painless! It would be unwise to be openly critical of your GP if your reason for wanting to change is dissatisfaction, especially if you wish to register with another GP in the same area. The problem may be effectively solved by seeing a different GP within the same practice if it is a large one and it is possible to do so; this would obviate the need to re-register. If this is not a viable option, however, there are basically two courses of action open to you. You can ask your present doctor to release you so that you can transfer to the GP of your choice (it is advisable to have checked with the new GP that he or she is willing to accept you on to his or her list). If both doctors agree – one to release you, the other to accept you – your transfer should be able to proceed immediately. The second, more lengthy, option is to apply for a transfer through the FPC, who will need to receive your medical card when you write giving the name of the doctor to whom you want to transfer. This option is usually considered by people who would rather not ask their GP directly for his consent to change.

Hospital treatment

Going into hospital can be upsetting, especially for the very young or the very old. It is a time when the support of family and friends is of paramount importance. Also important of course is the patient's relationship with the medical staff responsible for her treatment. The nurses looking after the patient are unlikely to be sympathetic if she is unappreciative and is reluctant to adapt to hospital routines. A patient cannot demand to be admitted to a particular hospital; nor has she the right to be treated by a particular doctor or surgeon. If you are not happy with the way you are being treated, or with your progress, you should express your feelings to your GP who should, if you request it, refer you to another consultant.

If you are in hospital for an operation, exploratory or otherwise, you will be asked to sign a consent form. When you give consent to an operation, make sure that you understand exactly what you are consenting to. Discuss things frankly with your surgeon, and, if there is something to which you do not agree, this can be written on the form. If there is any possibility that the operation will reveal the need for further urgent surgery, the surgeon should talk over any eventuality with you beforehand.

Private health

If you are considering opting for private treatment, there are a number of points to bear in mind. Perhaps the most important of these is that you should not allow the word 'private' to lead you into thinking that the treatment must necessarily be better than on the NHS – in reality there should be no difference in quality. The attraction of private medicine lies mainly in the fact that patients are not kept waiting either for appointments or for hospitalization. Appointments with a private practitioner can be arranged to suit the patient and tend to last longer than NHS consultations. Since for a fee private patients can have access to information not given out by the NHS (X-rays, reports etc.), there is a tendency once again to equate this with 'better' treatment.

Other advantages offered by private medicine lie in the provision of services that cannot be maintained on a large scale by the NHS. For example, regular screening, other than for cervical cancer, will only be carried out on women already showing symptoms of a particular illness, or with a strong family history of it. Mammography, for instance, is only carried out by the NHS on women in these categories. However, there are a number of private clinics that offer a variety of screening and other services: test-tube baby techniques, abortions, pregnancy testing, maternity care, cosmetic surgery etc.

In order to be able to afford private health care or treatment, it is essential for most people to have medical insurance (remember, though, that insurance does not cover 'elective' surgery, such as abortion or cosmetic surgery). This may be the first disadvantage. Other problems may arise from the fact that there is nothing to stop unscrupulous people posing as, say, acupuncturists or cosmetic surgeons without the necessary qualifications or experience. The first rule should always be to find out what the qualifications of the private practitioner are before agreeing to any form of treatment (unless he/she has been recommended by your GP). Since there are only a few NHS GPs practising alternative medicine – homeopathy, hypnosis and acupuncture – you may well have to look for a private practitioner if you are considering this form of treatment.

Cosmetic surgery is a growing industry. However, the NHS does not have the resources or expertise to carry out this largely unnecessary form of surgery on a wide scale. Cosmetic surgery for breast reduction or enlargement for example would only be done on the NHS if the doctor considered the woman's state of mind to be severely affected by her dissatisfaction with her breast size.

Making a complaint

NHS GPs

If you have a cause for complaint that discussion with your GP has been unable to sort out to your satisfaction and you wish to make that complaint formal, you should contact either the Family Practitioner Committee (FPC) or the General Medical Council (GMC). The FPC will investigate allegations concerning standards of service – diagnosis and treatment – and will decide whether there is a case to answer. The severest disciplinary action a GP can undergo may result in him/her being struck off the NHS register. Complaints concerning a GP's conduct are handled by the GMC. Again, if investigations show that the doctor was at fault, he/she may be prevented from practising within the NHS. Serious complaints of negligence or incompetence resulting in legal action can only be pursued through the courts. Be warned, however, that suing a doctor for negligence is not an easy process as the burden of proof lies with the patient. The few cases that do eventually come to court are usually lost by the patient so be sure you seek advice before pursuing legal action. It is also important to remember that financial compensation is usually only awarded if you suffer pain or inconvenience as a result of negligence.

Local Community Health Councils are a useful contact for advice if you wish to make a formal complaint against an NHS doctor. It is also important to remember that there are time limits within which certain complaints must be made, so check these out before putting anything in writing.

Hospitals

The complaints procedures in hospitals tend to be more complex and differ from one hospital to the next. However, the following general guidelines are a useful basis.

Preliminary complaints should be made to the staff treating you, the ward sister or the consultant. If you are not happy with the explanation given, and want to take the matter further, you should complain in writing to the Hospital Administrator, if it concerns something other than treatment, or the Regional Medical Officer, if it concerns a doctor's clinical judgement. Further action may be pursued if you are still dissatisfied. Complaints about a doctor's or nurse's conduct and allegations of negligence or incompetence should follow the same procedures as for GPs.

Private practitioners

One of the disadvantages of using a private practitioner is that it may be more difficult to pursue a complaint, since, more often than not, a dissatisfied patient is placed in the position of having to *prove* that the practitioner is in breach of contract. However, most professional bodies will look at complaints made against the members or practitioners on their register, whether they be private or NHS. Similarly, a private doctor can be sued for negligence or incompetence just as an NHS doctor can.

Useful Addresses

Please note that registered charities appreciate receiving an SAE with requests for information leaflets and material.

Active Birth Movement
32 Willow Road
London NW3
Tel: 01-794 5227

AIMS (Association for Improvements in Maternity Services)
163 Liverpool Road
London N1 0RF
Tel: 01-405 5195

British Association for Counselling
37a Sheep Lane
Rugby
Warwickshire
Tel: (0788) 78328

British Homeopathic Association
27a Devonshire Street
London W1N 1RJ
Tel: 01-935 2163

British Pregnancy Advisory Service
7 Belgrave Road
Victoria
London SW1
Tel: 01-222 0985

Brook Advisory Centres
233 Tottenham Court Road
London W1P 9AE
Tel: 01-580 2911/ 323 1522

Caesarean Support Group
7 Green Street
Willingham
Cambridgeshire

Criminal Injuries Compensation Board
Whitington House
19-30 Alfred Place
London WC1
Tel: 01-636 9501

Endometriosis Society
65 Holmdene Avenue
London SE14 9LD
Tel: 01-737 4764

Family Planning Association *and* Family Planning Information Service
27-35 Mortimer Street
London W1N 7RJ
Tel: 01-636 7866

Foresight
(for promoting preconceptual care)
Old Vicarage
Church Lane
Witley
Nr Godalming
Surrey

Gingerbread
(support for one-parent families)
35 Wellington Street
London WC2
Tel: 01-240 0953

Haemophilia Society
123 Westminster Bridge Road
London SE1 7HR
Tel: 01-928 2020

Health Education Council
78 New Oxford Street
London WC1A 1AH
Tel: 01-637 1881

Herpes Association
c/o Spare Rib
27 Clerkenwell Close
London EC1
Tel: 01-253 9792

Incontinence Advisory Service
The Disabled Living Foundation
380 Harrow Road
London W9 2HU
Tel: 01-289 6111

Institute for Complementary Medicine
21 Portland Place
London W1N 3AF
Tel: 01-636 9543

LaLeche League
(for promoting breast-feeding)
PO Box BM 3424
London WC1
Tel: 01-242 1278

Marie Stopes House
108 Whitfield Street
London W1P 6BE
Tel: 01-388 0662/2585

Marie Stopes Annexe
114 Whitfield Street
London W1P 5RW
Tel: 01-388 4843

Marie Stopes Centres

Leeds:
10 Queen Square
Leeds LS2 8AJ
Tel: (0532) 440685

Manchester:
1 Police Street
Manchester M2 7LQ
Tel: 061-832 4260

Mastectomy Association
26 Harrison Street
off Gray's Inn Road
London WC1H 8JG
Tel: 01-837 0908

Maternity Alliance
59-61 Camden High Street
London NW1 7JL
Tel: 01-388 6337

Miscarriage Association
18 Stoneybrook Close
West Bretton, Wakefield
West Yorkshire WF4 4TP
Tel: (0924) 85515

National Association for
the Childless
Birmingham Settlement
318 Summer Lane
Birmingham B19 3RL
Tel: 021-359 4887

National Association of
NFP Teachers
Birmingham Maternity
Unit, Queen Elizabeth
Medical Centre
Birmingham B15 2TG
Tel: 021-472 1377

National Association for
Pre-menstrual Syndrome
25 Market Street
Guildford
Surrey
Tel: (0483) 572715

National Childbirth Trust
9 Queensborough Terrace
London W2
Tel: 01-221 3833

National Osteoporosis
Society
Barton Meade House
PO Box 10
Radstock
Bath BA3 3YB
Tel: (0761) 32472

Natural Family
Planning Service
Catholic Marriage
Advisory Council
15 Lansdowne Road
London W11 3AJ
Tel: 01-727 0141

Patients Association
Room 33
18 Charing Cross Road
London WC2H 0HR
Tel: 01-240 0671

Pregnancy Advisory
Service
11-13 Charlotte Street
London W1P 1HD
Tel: 01-637 8962

Pre-menstrual Tension
Advisory Service
PO Box 268
Hove
East Sussex
Tel: (0273) 771366

Rape Crisis Centres

Belfast:
PO Box 46
Belfast BT2 7AR
Tel: (0232) 249696

Cardiff:
PO Box 18
108 Salisbury Road
Cathays
Cardiff
Tel: (0222) 373181

Dublin:
PO Box 1027
2 Lower Pembroke Street
Dublin 2
Tel: 0001-601470

Edinburgh:
PO Box 120, Head PO
Edinburgh EH1 3ND
Tel: 031-556 9437

London:
PO Box 69
London WC1X 9NJ
Tel: 01-278 3956/
837 1600

Society to Support Home
Confinements
17 Laburnum Avenue
Durham City DH1 4HA

Stitch Network
15 Matcham Road
London E11 3LE

Terrence Higgins Trust
BM AIDS
London WC1N 3XX
Tel: 01-278 8745
Helpline: 01-833 2971

Women's Health Concern
Ground Floor Flat
17 Earls Terrace
London W8 6LP
Tel: 01-602 6669

Women's National Cancer
Control Campaign
1 South Audley Street
London W1Y 5DQ
Tel: 01-499 7532/4

Women's Reproductive
Rights and Information
Centre (for information on
donor insemination)
52-54 Featherstone Street
London EC1
Tel: 01-251 6332

Bibliography

The A-Z of Women's Health, Derek Llewellyn-Jones, Oxford University Press/ Rainbird, 1983

Caring for Women's Health, Joan Jenkins, Women's Health Concern, 1985

Contraception: a practical and political guide, Rose Shapiro, Virago, 1987

Countdown to a Healthy Baby, Heather Bampfylde, Collins, 1984

Curing PMT the Drug-free Way, Moira Carpenter, Century, 1985

Dictionary of Pregnancy and Birth, Heather Welford, Allen & Unwin, 1986

Everywoman, Derek Llewellyn-Jones, Faber & Faber, 1986

Everywoman's Life Guide, Miriam Stoppard, Macdonald, 1985

Exercises for Childbirth, Barbara Dale & Johanna Roeber, Century, 1982

The Experience of Childbirth, Sheila Kitzinger, Pelican, 1967

Herpes, AIDS and Other Sexually Transmitted Diseases, Derek Llewellyn-Jones, Faber & Faber, 1985

How to Get Pregnant, Sherman J. Silber, Star Books, 1983

Hysterectomy: what it is and how to cope with it successfully, Suzie Hayman, Sheldon Press, 1986

The IUD: a woman's guide, Robert Snowden, Unwin Paperbacks, 1986

Let's Have Healthy Children, Adelle Davis, Unwin Paperbacks, 1986

Menopause, Raewyn Mackenzie, Sheldon Press, 1985

Miscarriage, Ann Oakley, Ann McPherson and Helen Roberts, Fontana, 1984

Natural Birth Control: a guide to contraception through fertility awareness, K. and J. Drake, Thorsons, 1984

The NEW Our Bodies Ourselves, Boston Health Collective, Simon & Schuster, 1984

Once a Month, Dr Katharina Dalton, Fontana, 1978

Pain-free Periods: natural ways to overcome menstrual problems, Stella Weller, Thorsons, 1986

A Patient's Guide to the National Health Service, Which?, Consumers' Association/Patients Association, 1983

The Pill, John Guillebaud, Oxford University Press, 1984

PMT: the unrecognized illness, Judy Lever, NEL Paperbacks, 1979

The PMT Solution: the nutritional approach, Dr Ann Nazzaro, Dr Donald Lombard with Dr David Horrobin, Adamarantine Press, 1985

The Pre-menstrual Syndrome: the curse that can be cured, Dr Caroline Shreeve, Thorsons, 1983

Understanding Pre-menstrual Tension, Dr Michael Brush, Pan Books, 1984

What Every Pregnant Woman Should Know: the truth about diet and drugs in pregnancy, Gail and Tom Brewer, Penguin Books, 1979

Woman's Experience of Sex, Sheila Kitzinger, Penguin Books, 1985

Women on Hysterectomy or How long before I can hang-glide?, Nikki Henriques and Anne Dickson, Thorsons, 1986

Women on Rape, Jane Dowdeswell, Thorsons, 1986

Index

Acknowledgements

Photographs
We would like to thank the following for supplying photographic material for use in the book:
Corry Bevington/photo co-op: page 73(right)
Gina Glover/photo co-op: page 73(bottom)
Maria Galland Cosmetics: title page
Marie Stopes Clinic: pages 9, 10, 11, 12 and 14
Mastectomy Association: page 141
Metropolitan Police Office: page 175(bottom)
National Osteoporosis Society: page 141
Weleda (UK) Ltd: page 130

Illustrations
The illustrations and diagrams throughout the book are by Jill and Phil Evans.

Our thanks also go to the Medical Research Council for allowing us to reproduce the drawing on page 24, and to the Office of Population Censuses and Surveys for supplying the data for the graph on page 76.